Absence of Chaos

by

Don Alexander

July 2016

Dedication:

To my only son, Dr. Blake Daniel
Alexander, in communication of that which
I understood not when I wallowed in the
hog pens of carnal ignorance

Chapter Outline:

Prologue

When considering the various "educated explanations" as to the origin of the universe and its life forms, the elementary definitions of imagination, opinion, hypothesis, theory, law of cause and effect, and established facts can be relied upon to separate truth from fiction. An opinion may be based on imagination or knowledge accumulated through observations concerning the foundation for the opinion or a combination thereof. Opinions generally contain some facts mixed with bias.

A hypothesis is a conclusion based

upon one or more observations colored by personal opinion and unsupported by scientific experiments which consistently yield the same result.

A theory is a hypothesis which is subject to being tested by repetitious and valid scientific experiments. An untested theory can never rise to the level of fact. An established fact is given birth by a theory which has withstood repetitive scientific testing yielding the same results conforming to the "law of cause and effect." A "happening," "state of being," or "event" is the effect; and the factor which gives birth to the effect is the cause.

The law of cause and effect states: any

factor in whose presence the effect fails to occur cannot be the cause; and conversely, any factor in whose absence the effect occurs cannot be the cause.

The big bang theory coupled with the evolution of all life forms descending from a single-celled living organism is a hypothesis and not a theory because there have been no scientific tests which produced repetitive and predictable results. The law of cause and effect cannot verify as factual an untested theory which is based only on biased and random observations mixed with imagination pertaining to a singularity (an event that has never been verified). The explanation for a singularity which violates the laws of physics (scientifically established facts) is highly unlikely and proclaiming such explanation to be factual is the height of intellectual dishonesty. From the beginning of human history upon Planet Earth, scientifically oriented individuals have pondered the origin of the sun, moon, stars, and Earth's life forms. The concept of spontaneous generation of primordial life forms and evolution of humans from lower life forms was proposed more than a thousand years before the birth of Charles Darwin along with the assumption that the celestial bodies are eternal without beginning or end.

Chapter one

Astounding complexity

It is not disputed by those with
doctorate degrees in physics, chemistry,
biology, genetics, astrophysics, theoretical
physics, quantum physics, mathematical
probabilities, and various other earth
sciences that darkness is the absence of
light; that chaos is the opposite of order;
that random chance is simple unplanned
happenstance rather than proceeding from
intelligent thought and creative design; that
any factor that is absent when the effect
occurs cannot be the cause; that any factor

that is present when the effect fails to occur cannot be the cause; that any event or happenstance with a mathematical probability of less than one chance in ten to the fiftieth power will never happen.

The molecular structure of what physicists refer to as light is actually pulsing waves of electromagnetic energy emitted by high temperature electrons falling back through quantum states toward the nucleus of individual atoms. The energy given off and measured in photons is equal to the difference in quantum state energy levels the electron traverses when falling toward ground state around the atom nucleus. The most common sources of electromagnetic energy in the universe are galaxies of stars which fuse hydrogen into helium through the heat of nuclear fusion. However, any combustible compound or hot metal composed of the elements is capable of generating light during combustion or temperature increase approaching the elemental melting point.

Chaotic happenstance can never produce symmetry and complexity peculiar to individual molecular structures such as snowflakes or human fingerprints or unique individual DNA.

Consider a late winter blizzard falling upon the land mass spanning Metropolitan St. Louis, Kansas City, Joplin and Springfield, Missouri (total of 14,407 square miles) from an altitude of 25,000 feet with a snowflake density of 120 flakes per minute per cubic foot over a period of three hours thereby encompassing more than two million trillion individual snowflakes. Then try to imagine all the snowfall around the entire globe for the last six thousand years with no two snowflakes having identical molecular structure.

Next, think about more than seven billion humans with ten fingers, ten toes, two palms two irises, a hundred trillion neuron synapses, and three billion base pairs of nucleic acid nucleotides in each of thirty-seven trillion individual body cells.

Scientists around the globe who study the human brain information processing capability estimate the bits of information processed per second at four hundred billion bits utilizing over a hundred billion brain neuron synapses interfacing in harmonious precision.

The internet encyclopedia Wikipedia published the following article concerning the complexity of the human autonomous nervous system. This excerpt is a minor portion of the overall publication:

"The **autonomic nervous system (ANS)** is a division of the <u>peripheral nervous system</u> that influences the function of internal organs.[1] The autonomic nervous system is a control system that acts largely unconsciously and regulates bodily functions such as the heart rate, digestion, respiratory rate, pupillary response, urination, and sexual arousal. This system is the primary mechanism in control of the fight-or-flight response and the freeze-and-dissociate response.[2] Within the brain, the autonomic nervous system is regulated by the hypothalamus. Autonomic functions include control of respiration, cardiac regulation (the cardiac control center), vasomotor activity (the vasomotor center), and certain reflex actions such as coughing, sneezing, swallowing and vomiting. Those are then subdivided into other areas and are also linked to ANS subsystems and nervous system external to the brain. The hypothalamus, just above the brain stem, acts as an integrator for autonomic functions, receiving ANS regulatory input from the limbic system to do so.[3]

The autonomic nervous system has two branches: the sympathetic nervous system and the parasympathetic nervous system.[4]The sympathetic nervous system is often considered the "fight or flight" system, while the parasympathetic nervous system is often considered the "rest and digest" or "feed and breed" system. In many cases, both of these systems have "opposite" actions where one system activates a physiological response and the other inhibits it. An older simplification of the sympathetic and parasympathetic nervous systems as "excitory" and "inhibitory" was overturned due to the many exceptions found. A more modern characterization is that the sympathetic nervous system is a "quick response mobilizing system" and the parasympathetic is a "more slowly activated dampening system", but even this has exceptions, such as in sexual arousal and orgasm, wherein both play a role.[3]

In general, the autonomic nervous system functions can be divided into sensory (afferent) and motor (efferent) subsystems. Within both, there are inhibitory and excitatory synapses between neurons. Relatively recently, a third subsystem of neurons that have been named non-noradrenergic, non-cholinergic transmitters (because they use nitric oxide as a neurotransmitter) have been described and found to be integral in autonomic function, in particular in the gut and the lungs.[5]

Although the ANS is also known as the visceral nervous system, the ANS is only connected with the motor side.[6] Most autonomous functions are involuntary but they can often work in conjunction with the <u>somatic nervous system</u> which provides voluntary control." (See actual internet article for contributor references).

The reason the foregoing publication is relevant to the issue of chaotic happenstance versus intelligent design is that random chance would be idiotic.

Based upon feedback from the Hubble space telescope astronomers now estimate that the observable universe contains around one hundred and seventy-six billion galaxies and single galaxies such as the Milky Way contain approximately one hundred billion stars (some galaxies are bigger than the Milky Way and some are smaller). The stars at the edge of the observable universe are estimated by astronomers to be 13.7 billion light years from Earth (one light year equals 186,200 times 60 times 60 times 24 times 365 which totals just under six trillion linear miles).

Considering the precise orbital balance of the solar system within the Milky Way, where indeed is the chaotic random chance happenstance that gave birth to the primordial elements within which is bound up all the matter, energy and motion within the know universe? Moreover, energy, matter and motion had to preexist any alleged "big bang" because the big banging could not have occurred in the absence of matter, energy and motion.

The late Sir Fred Hoyle, a British self-styled atheist, astronomer, scientist, mathematician, and physicist postulated that life was seeded in Earth from some-where in space but never had a clue as to where that panspermia life form originated. Published in his 1982/1984 books *Evolution from Space* (co-authored with Chandra Wickramasinghe), and partially incorporated into a Wikipedia article, Hoyle calculated that the chance of obtaining the required set of <u>enzymes</u> for even the simplest living cell without <u>panspermia</u> was one in 10 to the 40,000th power. Since the number of <u>atoms</u> in the known universe is infinitesimally tiny by comparison (10 to the 80th power), he argued that Earth as life's place of origin could be ruled out. He claimed:

> "The notion that not only the <u>biopolymer</u> but the operating program of a living cell could be arrived at by chance in a primor-dial organic soup here on the Earth is evidently nonsense of a

high order."

Though Hoyle had previously declared himself an atheist,[28] this apparent suggestion of a guiding hand led him to the conclusion that "a superintellect has monkeyed with physics, as well as with chemistry and biology, and ... there are no blind forces worth speaking about in nature."[29] He would go on to compare the random emergence of even the simplest cell without panspermia to the likelihood that "a tornado sweeping through a junk-yard might assemble a Boeing 747 from the materials therein" and to compare the chance of obtaining even a single functioning protein by chance combination

of <u>amino acids</u> to a solar system full of <u>blind</u>
men solving <u>Rubik's Cubes</u> simultaneously.
[30] Those who advocate the <u>intelligent design</u>
(ID) belief sometimes cite Hoyle's work in
this area to support the claim that the
universe was <u>fine tuned</u> in order to allow
intelligent life to be possible. Alfred Russel
of the <u>Uncommon Descent</u> community has
even gone so far as labeling Hoyle "an
atheist for ID".[31]

In tandem with other recognized
experts in mathematical probabilities Hoyle
further calculated that the chances of a
single-celled life form arising pursuant to
spontaneous generation is roughly one
chance in ten to the billionth power.

The 2004 Asian tsunami wave that killed roughly 230,000 people swept across the ocean at 310 to 620 mph rising to around 200 feet high upon approaching land. Suppose that the "fountains of the great deep" were broken up (as stated in Genesis, The Holy Bible) allowing flooding to cover the highest mountains on earth with tsunami waves racing across the oceans at the speed of sound. What type of sedimentation rock erosion would be anticipated? Consider this excerpt from the book "Deciding For Eternity" by Don Alexander:

"The "Geological Column" hard core evolutionists cling to is completely imaginary because petrified trees and skeletons of very large living creatures such as whales and dinosaurs have been found protruding through multiple sedimentary layers which the disciples of Darwin claim were laid down during geologic time windows spanning thousands and often millions of years between compilation of each layer of sediment (which solidified into rocks thereby providing a reliable yardstick to measure the age of our cosmos). These dedicated evolutionists date fossils according to location reflected on the geological column and *incredibly* also date the rock strata in accordance with the date assigned to fossils found therein. Regardless of such blatant circular reasoning, the geological column yardstick is now a companion supposition to the big bang. The unquestioned existence of "polystrate fossils" (fossils that protrude through two or more sedimentary rock layers) proves the so-called geological column to be nothing more than biased imagination. How could a petrified tree or whale skeleton protrude through multiple sedimentary layers if such layers were deposited through hydraulic actions over thousands or millions of years. How about petrified trees found standing upright (or leaning at an angle from upright) while protruding through multiple sedimentary layers? Such petrified trees most definitely were suddenly buried by devastating hydraulic action such as volcanic activities, extended flooding, etc.

Such examples of polystratic trees or polystratic animal fossils have been discovered at the Joggins mine fields in Nova Scotia; in a sandstone quarry near Edinburgh, Scotland; in Yellowstone National Park, USA; the Green River Formation in Wyoming; and in coal veins around the globe (which also disproves the premise that coal and oil deposits along with sedimentary rock layers were deposited by geologic forces over eons of time).

When both visible and tangible exceptions exist which directly contradict a hypothesis, the hypothesis is proven totally false in accordance with the law of cause and effect, elementary scientific reasoning, and plain common sense.

Some evolutionists state that life was seeded on earth by aliens from another planet. Others conclude that one or more prior universes imploded and gave birth to the present universe. Still others have imagined parallel universes which routinely share energy, matter and motion. Such mythical suppositions do not address the question of origin. What was the source of original energy, matter and motion? Simple logic dictates that since the universe is composed of energy, matter and motion, and since energy holds matter together and all motion is powered by energy, then there exists a ***creative source of energy*** that gave birth to the universe.

Nothingness by definition cannot create anything; neither by random chance nor by accidental design; and DNA coding precludes the concept of spontaneous generation of life forms which is an imaginary hypothesis rejected by competent scientists over a thousand years ago.

Evolutionists state that dinosaurs first appeared during the "Triassic geological period" approximately 231 million years ago and that these monstrous creatures reigned supreme upon earth from the start of the "Jurassic" period (135 to 200 million years ago) until dinosaurs became extinct around 66 million years ago. Dinosaurs were at the very top of the food chain and spent most of their lifetime hunting, feeding and producing offspring. The current human population doubling time cycle worldwide is 61 years. However, suppose that dinosaurs only doubled in population every million years over a time window of 165 million years. It is axiomatic then that dinosaurs would have completed 165 doubling time cycles. The number two doubled 165 times equals 4.676805239 raised to the 49th power which is the same as the integers 4,676,805,239 raised to the 40th

power such that 66 million years ago the

dinosaur population of earth would be

4,676,805,239,000,000,000,000,000

000,000,000,000,000,000,000,000, times

ten which tallies to forty billion, 676

million, 805 thousand, two hundred and

thirty-nine trillion, trillion, trillion times

one thousand which rounds off to forty

trillion, trillion, trillion, trillion.

Atheists and evolutionists in our public

schools teach our boys and girls that

humans evolved from monkeys or apes

around 200,000 year ago. Okay. Try this

basic math: If the human doubling time

cycle averaged one thousand years over

the past two hundred thousand years instead of the current human doubling time cycle of sixty-one years, then 200,000 divided by 1,000 =200 human doubling time cycles. Consequently, the number two raised to the 200th power equals a mere 1.606938044 raised to the 60th power which calculates to 160,693,804,400 to the 49th power which exceeds sixteen hundred billion trillion, trillion, trillion, trillion.

Since today's human population on earth is roughly seven billion, it does not require an intellectual giant to quickly determine that humans first appeared on Planet Earth approximately six thousand years ago.

What about the Grand Canyon and other rock strata which shows massive water erosion of sedimentary rock over eons of time? Well, natural wonders can be carved quickly by catastrophic hydraulic forces. Under water earthquakes or volcanic eruptions along the ocean floor faults can cause tsunami type tidal waves sweeping over water and land at velocities approaching the speed of sound and rising to tremendous heights. The weight of water is 8.34 pounds per gallon. Now, consider walls of water several hundred feet deep rising to six hundred feet or more and sweeping across rock strata at around 750 miles per hour. Would an unbiased individual rightly suspect that some massive erosion of rock strata might occur? Are there concrete examples of such rapid erosion? Yes. Indeed. Here are just a few:

The Little Grand Canyon of the Toutle River was created by a hot mud flow which diverted the Toutle River when Mt. St. Helens erupted. The mud flow carved a seventeen mile long series of canyons up to one hundred and forty feet deep within a few hours.

The Palouse Canyon in Eastern Washington which is three hundred to five hundred feet deep was eroded through solid basalt by Lake Missoula floods in two days.

Turbidity Currents are underground mud flows that later harden into rock strata. Thirty percent of all sedimentary rocks in the Grand Canyon is now believed by professors of geology to be simply turbidites. Some geologists further stipulate that perhaps fifty percent of global sedimentary rocks could be turbidites.

Another undeniable fact that precludes the fable of evolution of single-celled organisms into humans is the ***current global fossil record.*** Where are the billions of fossils that would have certainly fossilized world-wide demonstrating the claims of professors of evolution *if indeed* bacteria evolved into humans over *millions or billions of years*? The few alleged "missing links" hailed by evolutionists have now been proven to be completely *fraudulent or misidentified* by dishonest intellectuals craving public recognition.

Even if spontaneous generation was not chemically, genetically and mathematically impossible, where did the DNA code come from that would be absolutely necessary for the first spontaneously generated single-celled organism to reproduce itself through mitosis involving four reproductive phases known as prophase, metaphase, anaphase and telophase producing two new cell nuclei each of which contains a complete copy of parental chromosomes? To avoid immediate extinction, the first single-celled bacterium *must have begun* its life cycle with the coordinated capabilities to move, feed, digest nutrients to power all cellular activities, eliminate waste and reproduce itself. The simplest microbe known to modern scientists is "Mycoplasma genitalium" which contains 525 genes in its *genome* (complete set of DNA base pairs including genes). According to bio-engineers at Stanford University who conducted a simulation of a single reproductive cycle of this simplest *free-living bacterium*, the computerized model for one single reproductive division of the bacterium required **28** individually modeled and integrated subsystems using *128* computers running for 10 hours and generating one half *gigabyte* of data." (*end of excerpt*)

Incredibly large numbers are encountered within both the macro and the micro universe. The macro universe contains trillions of stars and interstellar masses whereas the micro universe numbers are even larger. For example, there are 602 billion trillion atomic mass units in one gram of matter. The DNA strands in one adult human body if unwound and laid out in a microscopic line would traverse 12 billion miles.

Chapter two

Intelligent perfection

An intelligent and unbiased review of the human brain, eyes, heart, lungs, and DNA coding should convince any rational human that evolution of humans from bacteria pursuant to trillions of minor random chance genetic mutations over billions of years is intellectually insulting.

According to an internet anatomy lesson by Tim Taylor, Anatomy and Physiology Instructor, posted on Wikipedia, the human brain weighs approximately three pounds and is made up of 75% water.

The brain's gray matter is made up of

neurons which capture and transmit brain

waves containing bits of information

translated by the interface between roughly

100 billion neurons and 500,000 billion

synapses. The brain's white matter consists

of dendrites and axons which create the

network by which neurons transmit bits of

information.

The brain contains approximately

100,000 miles of blood vessels. There are

no pain receptors in the brain which is

around 60% fat and continues to grow from

birth to age 18 with some additional growth

pursuant to concentrated brain activities.

The brain grows in the womb at the rate of

250,000 neurons per minute during the first trimester and is roughly adult sized at birth. The brain grows new neurons in response to mental activity.

The brain's capacity to experience emotions such as fear, joy, happiness, shyness, revulsion, etc. are developed in the womb. The sense of touch is developed during the eighth to twelfth weeks of pregnancy. The brain uses 20% of the total body's oxygen and 20% of the blood in the body and unconsciousness will occur if the supply of blood is lost for eight to ten seconds. Permanent brain damage or death will occur with blood circulation loss lasting from five to ten minutes. During waking hours the brain generates between 10 and 23 watts of power which would be enough to power a light bulb.

Language and consciousness lies within the function of the neocortex which makes up 76% of the brain's mass. The amygdala, a small area in the brain, allows the ability to read another person's face for clues as to what that individual is feeling with respect to emotions; and 50 to 70 % of physical ailments are due to psychological factors.

Internet research at multiple websites documenting human anatomy reveals that the human eye is marvelously complex and most certainly did not evolve random chance mutation piled upon more random chance happenstance over millions of years. Rather, a creative life force produced the human eye with all its unbelievable complexity and numerous component parts which must all co-exist simultaneously due to being connected to the optic nerve servicing a few million neuron synapses allowing the brain to convert electromagnetic energy waves into physical vision which discerns color, depth, distance and size.

The eye is an exceedingly complex two-piece unit. The smaller frontal unit called the cornea is linked to the larger unit called the sclera. The corneal segment is typically about 8mm in radius.

The iris and the pupil are seen instead of
the cornea due to the cornea's transpar-
ency. The fundus opposite the pupil
services the papilla and optic nerve fibers
connected to neuron synapses which
transmit precise information to the brain.
The human eye is equipped with three
layers of transparent structures: the
cornea and scalea; the choroid, ciliary
body and iris; and the retina. Within these
transparent structures reside the aqueous
humor, the vitreous body and the flexible
lens. The aqueous humor is a clear fluid
inside the anterior chamber (which also
contains the cornea, iris and exposed
lens) and the posterior chamber behind
the iris. A ligament made up of superfine

transparent fibers suspend the lens to the ciliary body.

A Wikipedia article pertaining to DNA coding refutes random chance DNA protein synthesis:

"Deoxyribonucleic acid or DNA is a nucleic acid that contains the genetic instructions used in the development and functioning of all known living organisms (with the exception of RNA viruses). The main role of DNA molecules is the long term storage of information. DNA is often compared to a set of blueprints, like a recipe or a code, since it contains the instructions needed to construct other components of cells, such as proteins and

RNA molecules. The DNA segments that carry this genetic information are called genes, but other DNA sequences have structural purposes, or are involved in regulating the use of this genetic information.

DNA consists of two long polymers of simple units called nucleotides, with backbones made of sugars and phosphate groups joined by ester bonds. These two strands run in opposite directions to each other and are therefore anti-parallel. Attached to each sugar is one of four types of molecules called bases. It is the sequence of these four bases along the backbone that encodes information. The information is read using the genetic code which specifies the sequence of amino acids within proteins. This code is read by copying stretches of DNA into the related nucleic acid RNA is a process called transcription.

Within cells, DNA is organized into long structures called chromosomes. These chromosomes are duplicated before cells divide, in a process called DNA replication. Animals, plants, fungi, and Eukaryotic organisms (protists) store most of their DNA inside the cell nucleus and some of their DNA in organelles, such as mitochondria or chloroplasts. In contrast, prokaryotes (bacteria and various archaea) store their DNA only in the cytoplasm. Within the chromosomes, chromatin proteins such as histones compact and organize DNA. These compact structures guide the interactions between DNA and other proteins, helping control which parts of the DNA are transcribed."

The above Wikipedia (internet encyclopedia) excerpt is included here to emphasize the fact that DNA is a genetic blueprint that cannot possibly evolve by random chance. One of the simplest single-celled life forms is the bacteria E. coli which has four million pairs of DNA nucleotides arranged in a specific sequence. The odds against mythical "mother nature" randomly evolving four million pairs of nucleotides in a precise sequence permitting E. coli to replicate itself can be compared to winning one million six-number lottery jackpots without missing a single number with respect to the sequence in which the winning lottery numbers are randomly selected. This is simply mathematically impossible.

Moreover, if single-celled organisms evolved into higher life forms, it would be axiomatic that simplicity must be continually evolving into complexity. This assertion is directly contrary to the established fact that left to random chance and the elements complexity always degenerates into simplicity tending toward chaos (the law of entropy). Evolutionists will chide that chicks randomly evolve from eggs. Does an egg hatch out over eons of time randomly without design, order or purpose in the absence of DNA coding? Or, does an egg hatch out in less than 30 days while demonstrating perfect design, order and purpose in total accord with its cellular DNA?

If a used Chevy is left exposed to the elements for a few million years, it will become a pile of chaotic elements rather than a better engineered car; and the contents of a junkyard are not likely to evolve into the space shuttle. On the other hand, a single living cell is far more complex than the space shuttle. It is very enlightening to consider the following statements by highly esteemed evolutionists, scientists, and professors:

"Natural selection acts only by taking advantage of slight successive variations; she can never take a great and sudden leap, but must advance by short and sure, though slow steps." (On the Origin of Species by Means of Natural Selection, or the Preservation of Favored Races in the Struggle for Life," 1859, p. 162; public domain). On page 158, Charles Darwin admitted:

"If it could be demonstrated that any complex organ existed, which could not possibly have been formed by numerous successive, slight modifications, my theory would absolutely break down."

Apparently, Darwin had a hard time swallowing his own theory. On page 155, he wrestled with his own conscience:

"To suppose that the eye with all its inimitable contrivances for adjusting the focus to different distances, for admitting different amounts of light, and for the correction of spherical and chromatic aberration, could have been formed by natural selection, seems, I freely confess, absurd in the highest degree."

"When it comes to the origin of life there are only two possibilities: creation or spontaneous generation. There is no third way. Spontaneous generation was disproved more than a hundred years ago, but that leads us to only one other conclusion, that of supernatural creation. We cannot accept that on philosophical grounds; therefore, we choose to believe the impossible: that life arose spontan- eously by chance." ("The Origin of Life," Scientific American, 191, P. 48, May, 1954)

"In the years after Darwin, his advocates hoped to find predictable progressions. In general, these have not been found -- yet the optimism has died hard, and some pure fantasy has crept into the textbooks." ("Evolution and the Fossil Record," Science, vol. 213, July, 1981, pg.

289)

"The pathetic thing is that we have scientists who are trying to prove evolution which no scientist can ever prove." (Nobel prize winning physicist Robert A. Millikan)

"The theory of evolution is one of the strangest phenomena of humanity; it is entirely destitute of proof." (World famous geologist from Canada, Sir William Dawson)

"The Darwinian theory of descent has not a single fact to confirm it in the realm of nature. It is not the result of scientific research, but purely the product of imagination. (Professor Fleischmann, University of Erlangen zoologist}

"There is not the slightest evidence that any of the major [animal] groups arose from any other." (Dr. Austin H. Clark, world famous American biologist)

"Darwin's theory of natural selection has never had any proof..." (Dr. Richard Goldschmidt, Professor of zoology, University of California)

"The Darwinian approach has consistently been to find some supporting fossil evidence, claim it as proof for evolution, and then ignore all the difficulties. It is, in fact, a common fantasy..." (Roger Lewin, science journalist).

Earth's fossil record is consistently

touted by big bang advocates and evolutionists as proof of the age of Earth and the evolution of bacteria into humans by evolving through marine, amphibian, reptilian, and mammalian life forms over the course of a few billion years. Actually, Earth's fossil record alone is sufficient to completely debunk Darwinian evolutionary concepts.

Human fossils have been found buried along with dinosaur remains. In Texas and the Dakotas, human tools and bones are found in the same fossil layer as dinosaur bones. Human footprints mingle with dinosaur and other mammal footprints in the same fossil layers in Texas and New

Mexico.

In Utah and Colorado, cliff and cave drawings depict dinosaur species dating between 400 and 1300 A.D. Decorated burial stones in Inca, Peru show various species of dinosaurs interacting with human figures dating between 500 and 1500 B.C. In Acambaro, Mexico, stone and ceramic figurines dated between 800 B.C. and 200 A.D. depict many species of dinosaurs. Since the first dinosaur fossils were not discovered until the nineteenth century, how would humans know about such monstrous reptiles hundreds of years earlier? According to big bang and Darwinian evolutionary claims, dinosaurs were blasted into extinction 65 million years ago by a little bang pursuant to a mass from space smacking into Earth; and humans did not pop out of monkeys until 200,000 years ago.

Approximately 95% of all Earth's fossil remains found to date are marine invertebrates; 4.74% are plant fossils; 0.25% are land invertebrates including insects; and 0.0125% are vertebrates (mostly fish). Over 90% of all vertebrate fossils discovered and recorded to date have consisted of less than one bone. Intermediate life forms only exist in text books promoting Darwinian evolutionary theory. Considering the abundance of species, both extinct and non-extinct, which all evolved according to the atheists and evolutionists from a single living cell randomly formed through time and chance over billions of years, the fossil record should be literally teaming with billions of intermediate life forms.

An interesting fossil did turn up which evolutionists were quick to claim as an intermediate life form. This fossil was named Archaeopteryx and hailed as a transition between reptiles and birds. Why? Well, it has teeth, and claws on its wings (it was later discovered to be a hoax). Great!! Now, where are the fossils of some intermediate life forms prior to and pursuant to Archaeopteryx? After all, there would have to be countless random mutations to transform a reptile into a bird.

"Piltdown man" (Eoanthropus Dawson) was disturbed from eternal rest in 1912. He was presented by the evolutionists who conveniently discovered him as a missing link between man and ape. More than 500 scientific essays were penned over four decades extolling the striking similarities between ape and man exemplified in the form of Piltdown man. He was a curious fossil. He consisted of two human skulls, an orangutan jaw, an elephant molar, a hippopotamus tooth, and a canine tooth from a chimpanzee. The skulls had been treated with acid, and the other "remains" were stained with an iron sulfate solution. The canine tooth was painted and the molars were filed down.

The orangutan jaw was modified to hide the fact that the jaw did not belong to a human skull. This concoction was strewn around a quarry in Piltdown, England for later discovery as the long awaited missing link. The individuals linked to the "discovery" were world famous evolutionists with impeccable credentials. Hence, no scientist bothered to closely examine Piltdown man for forty one years. When the shameless hoax was uncovered in 1953, Sir Kenneth Oakley found the human skulls to belong to

Ona Indians and the other remains were then properly identified as to origin.

"Ramapithecus" was widely acclaimed by evolutionary intellectuals as a "direct ancestor of humans." This fossil has been now identified as an extinct specimen of orangutans.

"Nebraska man" turned out to be a fraud based on a single tooth from a rare pig.

"Java man" consisted of pieces of bone, a skull cap and three teeth scattered over a wide area and dug up over twelve months. Today, we know the bones came from a human burial site; the femur is considered human; and the skull cap is believed to be from an ape.

"Neanderthal man" was promoted as a stooped ape-man. The fossil was eventually discovered to have been formed from a diseased primate.

"Australopithecus afarensis" or "Lucy" promised hope for a "missing link" find. Non-biased examination of Lucy's inner ear, skull and bones determined Lucy to be a pygmy chimpanzee with an upright stance.

"Homo erectus" fossils have been found through-out Earth. The fossils are human in origin and reflect individuals of small stature with proportionally smaller head and brain cavity, but within the range of people today. Middle ear studies show Homo erectus to be human. The fossils have been found in close proximity to other humans.

"Australopithecus africanus" and "Peking man" were hailed for years as true missing links but are now considered simply Homo erectus.

"Homo habilis" is now generally considered to be comprised of fragments from other fossils such as Homo erectus and Australopithecus.

"Toumai" is presented by promoters to be the "earliest member of the human family found thus far." A number of scientists examined the fossil and identified it as coming from an ape (October 2002).

Quibbling over an occasional faked or misidentified fossil is absolutely an exercise in mental nonsense. If evolutionary theory is anything more than a self-inflicted delusion, the fossil record would be vomiting intermediate life forms. The fossil record alone is sufficient to boot Darwinian evolutionary theories from our public schools.

A summation of more than one hundred articles covering the anatomy of the human heart is posted on Wikipedia:

"The heart is a muscular organ about the size of a closed fist that functions as the body's circulatory pump. It takes in deoxygenated blood through the veins and delivers it to the lungs for oxygenation before pumping it into the various arteries (which provide oxygen and nutrients to body tissues by transporting the blood throughout the body). The heart is located in the thoracic cavity medial to the lungs and posterior to the sternum.

On its superior end, the base of the heart is attached to the aorta,...pulmonary arteries and veins, and the vena cava. The inferior tip of the heart, known as the apex, rests just superior to the diaphragm.

The heart sits within a fluid-filled cavity called the pericardial cavity. The walls and lining of the pericardial cavity are a special membrane known as the pericardium. Pericardium is a type of serous membrane that produces serous fluid to lubricate the heart and prevent friction between the ever beating heart and its surrounding organs. Besides lubrication, the pericardium serves to hold the heart in position and maintain a hollow space for the heart to expand into when it is full. The pericardium has 2 layers—a visceral layer that covers the outside of the heart and a parietal layer that forms a sac around the outside of the pericardial cavity.

The heart wall is made of 3 layers: epicardium, myocardium and endocardium. Epicardium. The epicardium is the outer-most layer of the heart wall and is just another name for the visceral layer of the pericardium. Thus, the epicardium is a thin layer of serous membrane that helps to lubricate and protect the outside of the heart. Below the epicardium is the second, thicker layer of the heart wall: the myo-cardium.

The myocardium is the muscular middle layer of the heart wall that contains the cardiac muscle tissue. The myocardium makes up the majority of the thickness and mass of the heart wall and is the part of the heart responsible for pumping blood. Below the myocardium is the thin endo-cardium layer.

Endocardium is the simple squamous endothelium layer that lines the inside of the heart. The endocardium is very smooth and is responsible for keeping blood from sticking to the inside of the heart and forming potentially deadly blood clots.

The thickness of the heart wall varies in different parts of the heart. The atria of the heart have a very thin myocardium because they do not need to pump blood very far— only to the nearby ventricles. The ventricles, on the other hand, have a very thick myocardium to pump blood to the lungs or throughout the entire body.

The heart contains 4 chambers: the right atrium, left atrium, right ventricle, and left ventricle. The atria are smaller than the ventricles and have thinner, less muscular walls than the ventricles. The atria act as receiving chambers for blood, so they are connected to the veins that carry blood to the heart. The ventricles are the larger, stronger pumping chambers that send blood out of the heart. The ventricles are connected to the arteries that carry blood away from the heart.

The chambers on the right side of the heart are smaller and have less myocardium in their heart wall when compared to the left side of the heart. This difference in size between the sides of the heart is related to their functions and the size of the 2 circulatory loops. The right side of the heart maintains pulmonary circulation to the nearby lungs while the left side of the heart pumps blood all the way to the extremities of the body in the systemic circulatory loop.

The heart functions by pumping blood both to the lungs and to the systems of the body. To prevent blood from flowing backwards or "regurgitating" back into the heart, a system of one-way valves are present in the heart. The heart valves can be broken down into two types: atrioventricular and semilunar valves.

The atrioventricular (AV) valves are located in the middle of the heart between the atria and ventricles and only allow blood to flow from the atria into the ventricles. The AV valve on the right side of the heart is called the tricuspid because it is made of three cusps (flaps) that separate to allow blood to pass through and connect to block regurgitation of blood. The AV valve on the left side of the heart is called the mitral or the bicuspid valve because it has two cusps. The AV valves are attached on the ventricular side to tough strings called chordae tendineae. The chordae tendineae pull on the AV valves to keep them from folding backwards and allowing blood to regurgitate past them. During the contraction of the ventricles, the AV valves look like domed parachutes with the chordae tendineae acting as the ropes holding the parachutes taut.

The semilunar valves, so named for the crescent moon shape of their cusps, are located between the ventricles and the arteries that carry blood away from the heart. The semilunar valve on the right side of the heart is the pulmonary valve, so named because it prevents the backflow of blood from the pulmonary trunk into the right ventricle. The semilunar valve on the left side of the heart is theaortic valve, named for the fact that it prevents the aorta from regurgitating blood back into the left ventricle. The semilunar valves are smaller than the AV valves and do not have chordae tendineae to hold them in place. Instead, the cusps of the semilunar valves are cup shaped to catch regurgitating blood and use the blood's pressure to snap shut.

The heart is able to both set its own rhythm

and to conduct the signals necessary to maintain and coordinate this rhythm throughout its structures. About 1% of the cardiac muscle cells in the heart are responsible for forming the conduction system that sets the pace for the rest of the cardiac muscle cells.

The conduction system starts with the pacemaker of the heart—a small bundle of cells known as the sinoatrial (SA) node. The SA node is located in the wall of the right atrium inferior to the superior vena cava. The SA node is responsible for setting the pace of the heart as a whole and directly signals the atria to contract. The signal from the SA node is picked up by another mass of conductive tissue known as the atrioventricular (AV) node.

The AV node is located in the right atrium in the inferior portion of the interatrial septum. The AV node picks up the signal sent by the SA node and transmits it through the atrioventricular (AV) bundle. The AV bundle is a strand of conductive tissue that runs through the interatrial septum and into the interventricular septum. The AV bundle splits into left and right branches in the interventricular septum and continues running through the septum until they reach the apex of the heart. Branching off from the left and right bundle branches are many Purkinje fibers that carry the signal to the walls of the ventricles, stimulating the cardiac muscle cells to contract in a coordinated manner to efficiently pump blood out of the heart.

At any given time the chambers of the heart may found in one of two states: Systole. During systole, cardiac muscle tissue is contracting to push blood out of the chamber.

During diastole, the cardiac muscle cells relax to allow the chamber to fill with blood. Blood pressure increases in the major arteries during ventricular systole and decreases during ventricular diastole. This leads to the 2 numbers associated with blood pressure—systolic blood pressure is the higher number and diastolic blood pressure is the lower number. For example, a blood pressure of 120/80 describes the systolic pressure (120) and the diastolic pressure (80).

The cardiac cycle includes all of the events that take place during one heartbeat. There are 3 phases to the cardiac cycle: atrial systole, ventricular systole, and relaxation.

During the atrial systole phase of the cardiac cycle, the atria contract and push blood into the ventricles. To facilitate this filling, the AV valves stay open and the semilunar valves stay closed to keep arterial blood from re-entering the heart. The atria are much smaller than the ventricles, so they only fill about 25% of the ventricles during this phase. The ventricles remain in diastole during this phase.

During ventricular systole, the ventricles contract to push blood into the aorta and pulmonary trunk. The pressure of the ventricles forces the semilunar valves to open and the AV valves to close. This arrangement of valves allows for blood flow from the ventricles into the arteries. The cardiac muscles of the atria repolarize and enter the state of diastole during this phase.

During the relaxation phase, all 4 chambers of the heart are in diastole as blood pours into the heart from the veins. The ventricles fill to about 75% capacity during this phase and will be completely filled only after the atria enter systole. The cardiac muscle cells of the ventricles repolarize during this phase to prepare for the next round of depolarization and contraction. During this phase, the AV valves open to allow blood to flow freely into the ventricles while the semilunar valves close to prevent the regurgitation of blood from the great arteries into the ventricles.

Deoxygenated blood returning from the body first enters the heart from the superior and inferior vena cava. The blood enters the right atrium and is pumped through the tricuspid valve into the right ventricle. From the right ventricle, the blood is pumped through the pulmonary semilunar valve into the pulmonary trunk. The pulmonary trunk carries blood to the lungs where it releases carbon dioxide and absorbs oxygen. The blood in the lungs returns to the heart through the pulmonary veins. From the pulmonary veins, blood enters the heart again in the left atrium. The left atrium contracts to pump blood through the bicuspid (mitral) valve into the left ventricle. The left ventricle pumps blood through the aortic semilunar valve into the aorta. From the aorta, blood enters into systemic circulation throughout the body tissues until it returns to the heart via the vena cava and the cycle repeats."

In view of the foregoing anatomy

articles, to postulate that the physical bodies of fish, amphibians, reptiles, mammals and humans evolved from bacteria one random genetic mutation at a time over millions of generations is the height of intellectual dishonesty. And, it is also quite obvious that only an extremely biased atheist would contend that DNA coding resulted from billions of random chance protein synthesis cycles.

Chapter three

Soul trading

Mature atheists, agnostics, big bangers, Darwinian evolutionists, and those who believe in intelligent design and creation of the universe composed of nothing other than matter, energy and motion generally agree as to the elemental composition of what is perceived through the physical senses of vision, hearing, touching, tasting and smelling.

To fully comprehend that everything within the universe is made up of the primordial elements, it is necessary to consider the following consensus among 21st century scientists:

All matter within the universe is composed of what scientists call elements. The known elements total 112 of which 91 predate the appearance in space of our solar system. Currently, 84 of the 112 are considered "primordial" because they appear naturally and do not involve any laboratory synthesis.

The other 21 elements resulted from transmutation of some of the original elements through the process of radioactive decay, nuclear fission or nuclear fusion. Nothing but two dozen or so primordial elements are found in all organic and inorganic masses and are essential to the very existence of the physical structure of bacteria, plants, insects, animals and humans.

The known elements, including the nuclear elements resulting from nuclear fission and nuclear fusion, are:

hydrogen, beryllium, sodium, magnesium, lithium, potassium, calcium, scandium, titanium, vanadium, chromium, manganese, iron, cobalt, nickel, copper, zinc, boron, carbon, nitrogen, oxygen, fluorine, neon, aluminum, silicon, phosphorus, sulfur, chlorine, argon, gallium, germanium, arsenic, selenium,

bromine, krypton, rubidium, strontium,
yttrium, zirconium, niobium, molyb-
denum, technetium, ruthenium, rhodium,
palladium, silver, cadmium, indium, tin,
antimony, tellurium, iodine, xenon,
caesium, barium, hafnium, tantalum,
tungsten, rhenium, osmium, iridium,
platinum, gold, mercury, thallium, lead,
bismuth, polonium, astatine, radon,
francium, radium, rutherfordum, dubnium,
seaborgium, bothrium, hassium, meitnre-
ium, ununhexium, damstadium, roentger-
ium, copemicium, ununtrium, ununqua-
dum, ununpentium, ununsepium,
ununoctium, lanthanum, actinium, cerium,
thorium, protactinium, praseodymium,
neodymium, uranium, promethium,

neptunium, samarium, plutonium, europium, americium, gadolinium, curium, terbium, berkelium, dysprosium, californium, holmium, einsteinium, erbium, fermium, thulium, mendelevium, ytterbium, nobelium, lutetium, and lawrencium.

The physical molecular structure comprising the bodies of all life forms ranging from a single-celled bacterium to humans consists of the primordial elements bonded into molecular compounds by the binding energy within electrons. The thirty-seven trillion or so single cells the human body is composed of contain nothing but oxygen, carbon, hydrogen, nitrogen, calcium, phosphorus, potassium, sulfur, chlorine, sodium, magnesium, iron, cobalt, copper, zinc, iodine, selenium and fluorine, plus a few trace elements.

Electrons can be added, subtracted or shared between atoms of different elements to form molecules and molecules can be bonded together to form compounds such as my eyes, teeth, heart, and my other physical body parts. In fact, my physical body contains over 37 trillion individual cells and each cell is composed of the primordial elements in the form of atoms, molecules and compounds.

The molecular mix of chemical compounds making up a physical human body (to the nearest .01%) are 65% oxygen,

18% carbon, 10% hydrogen, 3% nitrogen, 1.5% calcium, 1.2% phosphorus, .2% potassium, .2% sulfur, .2% Chlorine, .1% sodium, .05% magnesium, <.05% iron, <.05% cobalt, <.05% copper, <.05% zinc, <.05% iodine, <.01% selenium, and <.01% fluorine, plus a few trace elements less than .01% each.

Although the primordial elements are composed entirely of energy, matter and motion, there is no life force bound up within the elements. The energy and motion within individual atoms of the elements results from the repelling and attracting force fields emitting from protons and electrons plus the strong and weak nuclear forces that keeps individual atoms from self-destructing. The life force originates outside of the physical body and is not subject to extinction upon the inability of the physical body to move, feed and reproduce. Since the individual's life force is not composed of the elements and the entire universe is indeed composed entirely of the elements, it is axiomatic that the life force of every living creature is definitely not some combination of energy, matter and motion.

Those who spend their short lives studying astrology, astronomy and astro-physics tell us that the sun represents 99.86% of the mass of our **solar system**.

The planets, comets, asteroids, and miscellaneous interstellar masses make up the other 0.14%.

The sun's diameter is estimated at 1,392,000 kilometers and its total mass is 330,000 times the mass of Earth. The sun orbits the Milky Way at a distance roughly equal to 25,000 light years with a velocity approaching 370 kilometers per second completing one orbit each 250,000,000 years. The chemical composition of the sun is 75% hydrogen, 23.31% helium and 1.69% oxygen, carbon, neon, and iron plus trace elements heavier than helium. This 1.69% of the sun's mass is 5,628 times the mass of Earth. The sun is just one of the

200 billion or more stars within Milky Way.

The Milky Way is one of an estimated 200 billion galaxies and its velocity is calculated at 550 to 600 kilometers per second. Milky Way's diameter is approximately 120,000 light years (one light year is just under ten trillion kilometers – the distance light travels in one year at a velocity of around 300,000 kilometers per second).

The number of individual stars within the known universe number into the trillions. The sun converts its hydrogen mass through nuclear fusion into helium at the rate of 620 million metric tons per second. The fused helium contains less mass than the converted hydrogen. The excess mass resulting from the fusion of hydrogen into helium radiates out from the sun in the form of pulsating waves of electromagnetic energy which we call sunlight.

With respect to origin of the elements, is it not obvious that if the universe is composed of the elements, then the origin of elements must have occurred outside of all space, distance and measured time relative to the known universe? That seems elementary. Would it not also be just as obvious that since the elements could not arise within our three dimensional universe there must exist another dimension or dimensions not perceivable within our recognized three dimensional habitat?

Doesn't that raise the question as to how many dimensions actually exist? And, if the elements sprang from perhaps seven dimensions, as Einstein believed, would not such dimensions have to be more complex than our perceived three dimensions (by virtue of such dimensions giving birth to the primordial elements)?

Is it not also quite feasible that some life force possessing creative power over the matter, energy and motion within our three dimensions most probably exists somewhere within perhaps seven dimensions?

The truth that the physical body and the individual life force are distinct and separable is recognized by roughly 85% of humanity. The other 14% are atheists and Darwinian evolutionists who knowingly, willingly, and without a singly shred of scientific evidence choose to believe that the physical bodies of living creatures and their inhabiting life forces are one and the same. Darwinian disciples and atheists seldom allow fully proven, established and documented scientific facts to get in the way of self-worship and egomania.

The human life force is referred to as the "human spirit" and the free will, desires and emotions combined with individual psychic drives and personal ambitions is considered to be the "human soul" which resides in the same spiritual realm as the spirit. The soul exercises free will and conceives individual actions and behavior to be carried out by the physical body to accommodate the lusts and desires of the soul. Within individual humans there is a continuous conflict between the lusts and desires of the soul and the moral consciousness of the spirit.

Within the spirit resides the innate knowledge as to morally acceptable human behavior. The individual does not need a moral counselor to condemn raping another man's wife and eating his children. Prior to

the Law of Moses and the code of
Hammurabi the most primitive human
society enforced individual rights
recognized by the local culture.

With the exception of atheistic
evolutionists humans believe in life after
physical death of their bodies within a
spiritual dimension where love, kindness,
charity, mercy and forgiveness during life
on earth are rewarded; and where hatred,
spite, pettiness, jealousy, greed, gluttony,
lying, murder, covetousness, stealing,
disrespect for parents and spousal infidelity
are punished. The modes of rewards and
punishment vary according to the belief
system adhered to as well as the duration

pertaining thereto.

Islam is the second largest religion on Earth (after Christianity) and was founded more than 500 years after the resurrection of Christ by an Arab named Muhammad who regarded himself as Allah's (Arabic for God's) guardian of the true faith of Abraham. Muhammad proclaimed that Jews and Christians distorted the revelations God gave to Abraham, Moses, Jesus and other prophets by text altering and misinterpretation. The sacred writings of Islam are referred to as the Qur'an (God's revelations to Muhammad) and the Sunnah (words and deeds of Muhammad, God's final prophet to Earth). Followers of Islam are known as Muslims who deny that God had a son. They believe that Jesus was a mere prophet, that he escaped into Paradise and did not sacrifice himself for the sins of all humans. Everyone must save themselves from hell by accumulating more good works than evil works, and true believers may have to spend some time in hell to atone for insufficient good works compared to evil works. Upon birth, an individual's record is opened in Paradise and the individual becomes chargeable upon reaching the age of accountability (puberty). It is permissible to lie, steal, kill, rob, rape and pillage in the service of Allah (converting infidels from the error of their ways). Paradise is a place

of feasting, drinking, and sexual gratification surrounded by a swarm of virgins.
Muslims who die or commit suicide in
service to Allah are ushered directly into
Paradise. Muslims condemn homosexuality, adultery, eating pork and gambling.

Zoroastrianism is a religion with
approximately 200,000 followers which
preaches "save yourself" through good
thoughts, good words, and good deeds.
Zoroastrianism is also referred to as
Mazdaism and first appears in recorded
history during the seventh century BC.
Zoroaster is the main prophet for the belief
system and invented the term "Ahura
Mazda" to name the one universal and

transcendental god. Mazdaism embraces the concepts of good and evil. There exists an immortal adversary of Ahura Mazda dedicated to evil (Druj). Humanity is drawn into the conflict wherein Ahura Mazda is ultimately victorious and time ends with the renovation of the universe. Thereafter, all creation is reunited in Ahura Mazda. The collection of sacred texts are called "Avesta." Good thoughts, words and deeds are required to ensure happiness and ward off chaos. Today, followers of Zoroastrianism are located primarily in Iran, India and Pakistan.

Unitarian Universalists promote world unity and the inherent goodness of humans. All religions are embraced and are considered to be of equal merit. The ultimate achievement is religious unity and a single world government. Peace through good thoughts, good deeds and unbiased tolerance is the main tenet. Otherwise, all followers are encouraged to worship as they choose. Consequently, Unitarian Universalists are everything to everybody and humans will eventually live together happily as quickly as unity becomes reality. Abortion, homosexuality, and same sex marriage are smiled upon.

A fiction writer, L. Ron Hubbard, gave birth in 1960 to what has become known as Scientology and described as the

study and handling of the human spirit in
relationship to itself, others, and all life.
The sacred texts are various books written
by Hubbard. Man (a gender neutral term to
encompass all humans) is basically good
but life experiences lead him into evil. Man
errs by trying to solve his problems from
his own point of view rather than achieving
greater spiritual awareness through
learning, auditing and training. Man is a
spiritual being whose existence spans more
than one lifetime. Man is endowed with
abilities well beyond those he normally
considers he possesses. What is true for
man is what he has observed to be true.
During each reincarnation, man applies the
knowledge and increased spiritual

awareness he acquired during the previous life.

Man can improve his quality of life to the degree he continues to preserve his spiritual integrity and remains honest and decent thus achieving certainty of spiritual existence and a relationship with whatever supreme being he believes exists outside of himself. Scientology organizations provide ongoing auditing and counseling. Because man alone controls his earthly and eternal existence, Scientology is a somewhat unusual "save yourself" belief system with Hubbard's home spun psychiatry and hypnosis woven within the pseudo-scientific orientation.

The Bahai followers believe in one god who created everything; but the Bahai god is transcendent and unknowable who has and will continue to send great prophets to humanity through which the unknown deity has revealed a series of messages. Bahai prophets thus far have been Adam, Abraham, Moses, Krishna, Zoroaster, Buddha, Jesus Christ, Mohammed, The Bab, and Bahaullah. Another prophet is not expected for many centuries into the future. Bahai teaches the essential unity of the great world religions as arising from the same spiritual source but splintered by conditions at the time of founding and by accretions following the death of the founder.

The Bahai faithful believe that all individuals possess an immortal soul not subject to decomposition. At death, the soul is freed to travel throughout the spiritual universe which is a timeless and placeless extension of the known universe. The sacred texts are a collection of the writings of Abdul-Baha, The Bab, and Bahaullah plus miscellaneous Bahai scriptures which were first circulated during the nineteenth century AD.

Bahai further teaches that the happiness of mankind as well as world peace and security are unattainable until global unity is firmly established. Bahai promotes gender and race equality, world government, freedom of expression and assembly, world peace, religious tolerance, and religious cooperation. Bahai rejects homosexuality while calling for equal dignity and respect for all peoples, the elimination of poverty and excessive wealth, universal education and economic justice. Mankind must control the present and future through unity and global cooperation.

The sacred texts of Hinduism are collectively referred to as "the Vedas" and the written forms date between 600 to 300 BC. Hinduism views the entire universe as one divine entity who is at one with the universe but transcends it as well. Brahma is the creator who is always creating new realities. Dharma is the eternal order, religion, law and duty.

Vishnu preserves the creations of Brahma and travels between heaven and earth in one of ten incarnations. Shiva is the destroyer of eternal order but can be compassionate and erotic. Hinduism is splintered into various groupings which worship local gods and goddesses. The two major divisions of Hinduism are Vaishnavaism (Vishnu is the ultimate deity) and Shivaism (Shiva is the ultimate deity). The main tenet of Hinduism is the transmigration of the soul -- a continuous cycle of birth, life, death, and rebirth through many lifetimes (referred to as "samsara"). Hindu priests serve at rituals and worship ceremonies but are considered unnecessary in rural areas where priestly duties are carried out by local non-Brahmins.

The four aims of Hinduism or "the doctrine of the fourfold end of life" are Dharma (religious righteousness), Artha (economic success and wealth), Kama (gratification of the senses such as sex, pleasure, and mental enjoyment), and Moska (liberation from samsara). The three goals of the "pravritti" (those who are in the world) are Dharma, Artha, and Kama. The main goal for the "nivritta" (those who renounce the world) is Moska. Liberation from samsara, thus becoming one with the universe, is the supreme goal of mankind.

Kama also refers to the accumulation of an individual's good and bad deeds.

An overload of bad Kama might result in rebirth as an animal or insect. The unequal distribution of wealth, prestige, and suffering is believed to be the result of one's previous acts during the current life and previous lives. Meditation is practiced with Yoga being the most observed. Other Hindu activities include rituals, daily prayers, and ceremonial dinners for various deities.

Hinduism is a "save yourself" belief

system where good thoughts, good
intentions, and good deeds will ultimately
be rewarded with cessation of
reincarnation.

Buddhism was founded during the
second half of the sixth century BC by
Siddhartha Gautama, son of King Gautama
who ruled over a small district in the
Himalayas between India and Nepal. As a
young man, Siddhartha wandered outside
the palace and observed a leper, a corpse,
and an ascetic whereupon he concluded that
happiness is an illusion. After he fathered a
son to ensure the royal bloodline, Siddhartha began a pilgrimage of inquiry and
asceticism wherein he was influenced by

two Brahmin hermits and later by five monks.

After years of seeking communion with the supreme cosmic spirit, Siddhartha claimed to have discovered the four noble truths (Pativedhanana) and pronounced himself the Buddha. He labored some forty years spreading the Buddha doctrines and died a questionable death at age 80 (it is reported that he was poisoned by a blacksmith). The teachings of Buddha are referred to as Dharma. Following Siddhartha's death, his followers convened to create tenets they could all accept within the caste system which required a series of rebirths to move up through the system.

The Buddha rejected the concepts of a supreme being and eternal souls. Whatever gods inhabit the cosmos are impermanent and are reincarnated like humans. The cessation of rebirths is named "nirvana" wherein the individual being becomes one with the Universal Soul. Nirvana is the ultimate achievement.

Karma (tally of good and bad deeds) determines the kind of rebirth and quality of life after rebirth. The path to nirvana is to follow the four noble truths -- the universality of suffering; the origin of suffering; overcoming of suffering; and the suppression of suffering. Lustful desires cause suffering which is experienced

during rebirth, aging, death and rebirth.
Suffering can be overcome by suppression
of the desires causing one to suffer. The
way leading to suppression of suffering is a
noble path with eight branches -- right
aspirations, right speech, right conduct,
right livelihood, right effort, right
concentration, right views of understanding, and right mindfulness. The eight
branches are different dimensions of a total
way of life.

Several lives are required to achieve
nirvana. The journey is long and difficult
with inner peace and harmony as one
approaches nirvana; then nothingness. The
sacred texts of Buddhism were compiled

around 80 BC and are referred to as the Pali
Canon (also called the Tripitaka). In
summary, Buddhism rejects the concept of
a supreme being and teaches that human
works are disciplined by cycles of
reincarnation.

Jainism, a heretical movement within
Hinduism, was founded by a man named
Mahavira. The sacred texts of Jainism are
the twelve "angas" plus lesser writings
which appeared in written form around
161600 AD. At age 30, it is believed by
followers of Jainism that Mahavira decided
to live a life of self-denial and wandered
naked through India for twelve years before
achieving "enlightenment." In his thirteenth

year of naked wandering, in a squatting position exposed to the sun with his knees high and his head low, in deep meditation, Mahavira reached nirvana whereupon he stopped living by himself and attracted disciples. He preached his revelations until his death at which time he allegedly boasted of 14,000 monks within his brotherhood.

Although Mahavira was steadfastly opposed to the concept of God or gods, his followers elevated him to deity claiming that he descended from heaven without sin and having all knowledge. Jainism preaches self denial as the path to nirvana. The Five Great Vows renounce killing any living thing; lying, greed, sexual pleasure, and worldly attachments. Monks were taught to avoid women entirely because Mahavira believed they were the cause of all types of evil.

Taoism traces its roots to Lao-Tzu around 600 BC. The sacred text is Tao-te-Ching (also known as Daodejing). Taoism was originally a hodgepodge of psychology and philosophy that became a religion in 440 AD and benefited from state support until the fall of the Ching Dynasty in 1911 AD. Today, Taoism has roughly twenty million followers centered mainly in Taiwan off the mainland of China.

Taoism rejects the concept of a

personalized deity. Tao is the life force which flows through the universe and all life. The ultimate achievement is to harmonize with Tao. Taoists believe in letting nature take its course unimpeded by mankind. Time is cyclical and not linear. Kindness is always reciprocated. Left to themselves, people will be compassionate without expecting a reward. Tao regulates and balances natural processes and embodies the harmony of opposites -- no love without hate; no light without dark; no male without female, etc. There is no God to hear prayers or to act on them. Taoists seek answers to life's questions through inner meditation and outer observation.

Evolution is the religion of atheists who deny that any supreme being exits; and that all life forms evolved from a single living cell which created itself through a chain of unrelated and purely random events. From this original living cell evolved every life form that exists today or has ever existed in the past eons of time. The elements within the universe emerged from nothingness by random chance for no reason and without purpose. There is no life of any kind following physical death. Humanity is simply an accidental life form produced by natural selection and survival of the fittest.

Judaism is a belief system traced

back to a man called Abraham who lived in
Ur of the Chaldees within the fertile
crescent around 1900 BC. Although his
family worshiped idols, Abraham believed
there is an unseen supreme being who
created the universe and all life forms.
According to Judaism, Abraham meditated
upon and prayed to the invisible God until
God called him out from among the idol
worshipers and instructed him to travel to a
land which God would give to him and to
his seed. The "promised land" would be
shown to him as he traveled. Abraham
obeyed God and went out not knowing
where he was going. God led him to the
land of Canaan where Abraham lived as a
shepherd and fathered Ishmael and Isaac.

Ishmael was the firstborn, but his mother was a bond-servant. Isaac's mother was Sarah, Abraham's wife. Ishmael and his seed fathered the Arabs; and Isaac's descendants became known as Jews.

Isaac's wife birthed two sons named Jacob and Esau. Jacob's name was changed to Israel and his twelve sons begat twelve tribes who collectively are referred to as "the children of Israel." Esau's descendants are called Edomites. Esau, being the firstborn, sold his birthright to Jacob for a bowl of stew.

The sacred texts of Judaism are the thirty-nine narratives which make up the total writings within the Old Testament of

the Holy Bible. Followers of Judaism
believe in one true God, sin and
righteousness, resurrection from the dead,
heaven and hell, Satan and the angels,
eternal life, and eternal punishment. They
further believe in sacrificial offerings
(animal sacrifices) to obtain forgiveness for
breaking God's law. Moses, the greatest of
God's prophets, received God's law while
seeking God on Mount Sinai around 1491
BC. The Law of Moses covers criminal,
civil and religious law. Religious law
involves a priesthood, rituals, sacrifices,
holy days, annual feasts, an annual Day of
Atonement, a Sabbath day (every 7th day),
the Sabbath year (every 7th year), and the

year of Jubilee (every 50th year wherein all bond servants are freed and all real estate reverts to the original tribal family).

The ten main points of the Law of Moses are referred to as God's Ten Commandments -- worship God only; do not set up nor worship any graven image; do not speak of God in an irreverent manner; remember and keep the Sabbath day; do not commit adultery; do not steal; do not murder; do not bear false witness; do not covet; and honor both father and mother. The thirty-nine Old Testament narratives were written and compiled between 1491 and 397 BC by a total of thirty-one different authors. The first five narratives are ascribed to Moses and are referred to as the "Pentateuch" (Genesis, Exodus, Leviticus, Numbers and Deuteronomy).

Followers of Judaism who rejected the teachings of Jesus Christ are still waiting for their prophesied Messiah who the Hebrew prophets said would restore Israel to the glory the nation enjoyed under King David and would make Jerusalem the center of world government.

Jews (children of Israel) have been the most hated and persecuted of all nationalities (antisemitism) since 825 BC and continuing to the present time.

Following approximately 2,000 years

of dispersion among Gentile nations, the Jews began returning to their homeland pursuant to a United Nations mandate issued in 1948 AD. During the same year, the tiny nation of Israel declared its independence. The nation of Israel has remained independent and has become one of the worlds most lethal military powers.

With reference to all world religions other than Judaism and Christianity, there is not a single fact to support the various belief systems. Reincarnation was plucked from the imagination of humans. There has never been a documented case of reincar- nation. The basic goodness of mankind has been repeatedly proven to be an illusion along with global unity. Good thoughts, good words and good deeds may be exhibited from time to time due to will worship and voluntary self-denial, but do not represent routine human behavior. Mankind generally exhibits greed, selfishness, envy, cruelty, and hatred for the socially outcast.

Past and present wars are very obvious examples of true human nature. The Law of Cause and Effect eliminates reincarnation as being factual. The basic goodness of humanity has never been demonstrated within relationships between the races, nor between general populations, nor between nations; and

very seldom between individuals.
Bastardized versions of Christianity, such
as Islam, seek to justify greed, lust. and
violence.

Although Judaism is Christianity
concealed; and Christianity is Judaism
revealed, followers of Judaism still look
to personal works and animal sacrifices
as the ticket to eternal life with God.
Only Christians believe and teach that
redemption from personal sins and
eternal life in Heaven flow solely from
faith in and acceptance of a divine
sacrifice provided by God; and that the
very best of human works fall short of the
standards reflected in God's law. Prior to
the law of Moses, sin was not charged to

individuals because in the absence of law there can be no sin. God gave the law to demonstrate to every human the need for a divine sacrifice. Jesus said he is that sacrifice; and that the souls of the dead prior to his offering of himself before God were confined to either Paradise or Hell awaiting his coming.

The ten commandments on the surface appear easy to keep yet the Bible declares that no man except Jesus ever kept God's laws. The reason that humans cannot keep God's law is because sin originates in our thoughts and does not depend upon execution of our mental rebellion and lust. We conceive sin in our mind and when we act upon sinful thoughts we incur the consequences of our acts. However, because sin originates in our thoughts, humans are not capable of living a sinless life. Therefore, we can only be redeemed back to God by a divine sacrificial lamb (Jesus Christ).

The roots of Christianity within Judaism and the teachings of Jesus Christ are found within the pages of the Holy Bible. Those who reject Jesus as the Son of God have labored for twenty centuries to disprove Biblical Scriptures. To date, the enemies of Jesus have failed to come up with a single proven example of Biblical error. On the other hand, the Holy Scriptures repeatedly demonstrate the unerring accuracy found therein.

There are Biblical statements which cannot be proven to be true, but neither can they be proven to be false. That is why the Scriptures declare that the redemption of human souls results from grace through faith in Jesus Christ and not through that which is yet to be revealed.

Because no religion nor belief system, other than Christianity, predicates eternal life with God in Heaven as God's "free gift to humanity," it is more than prudent to consider what the Holy Bible reveals to mankind concerning life, death, time and eternity. Jesus Christ, through the works he performed, has already demonstrated to any logical and rational person that God exists. If, then, an individual accepts God's existence, the acts attributed to God as described within the Holy Bible are insignificant displays of his unlimited power. Just one glimpse at the atomic structure of atoms or the awesome design obvious within DNA indicates the unfathomable intelligence and preexistence of our creator. The question is not whether the described events occurred. The question is -- does God exist?

It is most advisable to avoid the "killing Lazarus" mindset when reading God's revelation to humanity. During Jesus' earthly ministry He raised a man named Lazarus from the dead before

numerous eyewitnesses. Lazarus had
been dead for four days and was already
decomposing. The enemies of Jesus
decided the only way to deal with such a
miracle was to kill Lazarus thereby
destroying the evidence of his
resurrection.

Jesus Christ is the only "God-man"
who proved His divinity by His personal
works whereby He raised individuals
from the dead; healed the sick and
handicapped within entire cities; fed
9,000 men plus women and children with
a handful of bread and fish; cast out evil
spirits; opened deaf ears and blind eyes;
walked upon the raging sea; turned water
into wine; willingly offered up His own

body and blood as humanity's sacrificial lamb to redeem fallen humans who believed in and accepted His sacrifice; was Himself raised from the dead and taught His apostles and disciples for forty days following His resurrection from the dead; and was personally seen by over 500 people at one location before ascending back to heaven from whence He came to minister to humanity.

The enemies of Jesus never disputed His miracles but rather simply said He raised the dead and cast out demons by the power of Satan while committing blasphemy. Jesus taught and proved that there is one true God willing to accept His death as a sacrificial lamb

to redeem humanity back to a state of
innocence while embracing all believers
as "our Heavenly Father."

Along the pathway of mortal life
on earth, at a time not of our choosing,
our physical bodies have an appointment
with death which frees our soul and spirit
to experience eternal life within a new
and glorified body or to suffer eternal
damnation and punishment for refusing to
believe in and accept Jesus Christ as our
"divine sacrificial lamb" thereby "dying
in an eternal state of personal sin against
God."

Jesus asked: What shall it profit a
man if he should gain the whole world
and lose his own soul; or what will a man

give in exchange for his soul." It is sad but true that 80% of humanity trade their souls for status or wealth or other lusts of the flesh that are temporary and can never satisfy the spirit or soul. It is well said that sin will cost you more than you want to pay; take you farther than you want to go; and keep you longer than you want to stay. From eternity, God has said: "I set before you life and death. Choose life."

Chapter four

Torn veil

It is testified to by numerous eyewitnesses that a very remarkable event took place around 33 AD. A humble teacher without any formal education put to silence the most educated and revered members of Israeli society. He and a dozen of his most loyal companions went to the tomb of a man (called Lazarus) who had provided them with food and lodging when they passed by his home. The deceased had a couple of sisters who mourned his passing and were upset because the teacher (called

Jesus of Nazareth) did not arrive until after their brother's funeral. Jesus assured the sisters (Mary and Martha) that he shared their sorrow and would raise their brother from the dead. Martha was astonished and exclaimed that Lazarus had been dead for four days and that his body was already rotting. Jesus told her that he possessed the power to resurrect and was going to raise Lazarus from his grave. Jesus prayed a short prayer in which he referred to God as his father and then he called Lazarus forth from among the dead. Lazarus appeared in the mouth of his tomb. He was wrapped head to toe in grave clothes. Jesus told the mourners to unwind Lazarus and let him go.

Needless to say, this resurrection caused a great clamor among the immediate community. The Israeli religious leaders could not deny that a great miracle had indeed taken place. They conferred among themselves and determined that the best way to keep other Jews from following Jesus was to kill Lazarus thereby destroying the physical evidence of his resurrection.

Now, consider for a moment this historical event witnessed by the followers of Jesus and a considerable crowd of friends and relatives who had come to mourn at the tomb. Even the Chief Priests and lawyers who hated Jesus did not deny the resurrection of Lazarus. That would have been useless in the face of so many eyewitnesses. It seemed more prudent for them to simply murder Lazarus along with Jesus. Such is the mental state referred to as "killing Lazarus." It is the mindset to deny that which is patently obvious to any rational person and the willingness to commit any act including cold blooded murder to perpetuate status and tradition.

Such a mindset is even more incomprehensible when it is freely exhibited by those vested with the power to judge others and to teach moral law. Nevertheless, just as we accept the historical fact that the Germanic hordes sacked Rome in the fourth century AD, we also know that the religious rulers of the Jews in 34 AD proclaimed Jesus of Nazareth to be a fraud, a liar, a blasphemer, demon-possessed, a false teacher, and worthy to be crucified.

Wherever Jesus traveled and preached some or all of his twelve chosen companions were present with him as well as other eyewitnesses (often numbering into the thousands) who watched him perform acts which mortal men cannot perform under any circumstances. Even the most vicious and most murderous of the enemies of Jesus did not deny that he performed these miracles.

Jesus preached a notable sermon wherein he said that the kingdom of heaven belongs to the poor in spirit; that those who mourn will be comforted; that the meek shall inherit the earth; that those who hunger and thirst after righteousness will be filled; that the merciful shall obtain mercy; that the pure in heart shall see God; that the peacemakers shall be called the children of God; and that those who are reviled and persecuted for seeking after righteousness will be blessed.

While being followed by a multitude of people, Jesus stopped and healed a leper. He was shortly thereafter met by two individuals possessed by demons and the demons recognized Jesus and asked him if he had come to torment them before their

time. Jesus cast out the demons and
continued on his way.

Later, Jesus was with some of his
disciples in a boat when a great storm
threatened to capsize the vessel. Jesus
rebuked the raging storm and immediately
the sea became calm. Eyewitnesses to the
miracles performed by Jesus reported the
events to others and huge crowds followed
after Jesus and were present when he
restored sight to the blind, healed crippled
and withered limbs, cast out demons, and
fed thousands of people with a small boy's
lunch. These curious crowds further
directly observed Jesus heal people with
every other type of physical disability and

disease. Are these the acts of a liar, fraud and blasphemer? Remember, we are speaking of direct, eyewitness testimony by a multitude of people who came out to see and hear Jesus. The opposition did not deny that Jesus wrought such miracles as reported by eyewitnesses. Rather, they decided their best option to rid themselves of Jesus and his influence would be to find a way to kill him in secret or to use false accusers to convince the Roman governor, Pontius Pilate, that Jesus should be executed.

The religious leaders (priests, lawyers and scribes) who wanted Jesus dead knew that two basic strategies were worth pursuing should they fail to kill him themselves. If Jesus could be publicly heard to preach contrary to the Law of Moses, the common people would no doubt stone him to death. On the other hand, if Jesus could be tricked into saying something contrary to the interests of Rome, they could drag him before Pilate hoping for a death sentence. Several attempts were made unsuccessfully to carry out both strategies.

A woman was dragged from a bed of adultery and paraded before Jesus. The clever conspirators demanded to know whether Jesus (contrary to his teaching of forgiveness) would say that the woman

should be stoned according to the law of
Moses. If Jesus answered no, he, himself,
would probably be stoned. If he answered
yes, he would appear two-faced because he
preached forgiveness. Jesus answered that
whoever was without sin among the
woman's accusers should cast the first
stone.

The tricksters asked Jesus whether he
believed it was proper for Jews to pay
tribute to the Romans. If Jesus answered
yes, the common people would turn on him
because they loathed the Roman taxes. If he
answered no, he could be accused before
Pilate and probably crucified. Jesus held up
a coin and asked whose image and super-

scription were engraved upon it. When the audience answered that the image repre-sented Caesar, Jesus advised them to render unto Caesar the things which pertain to Caesar and unto God the things belonging to God.

The Sadducees did not believe in resurrection after death and they came to Jesus hoping to make his teaching affirming the resurrection appear foolish to the masses. Moses had commanded that if a Hebrew man died childless his next of kin should take the widow and raise up seed unto the deceased so that his name would not perish for lack of offspring. The Sadducees told Jesus that a man had taken a wife and died childless. The dead man had six brothers. The eldest brother took the widow as his wife, but he also died childless. Then, the remaining brothers in turn took the widow as a wife, but they also died childless. Finally, the woman died. Thus, inquired the Sadducees, whose wife would the woman be in the resurrection since all seven brothers had taken her for a wife. Jesus replied that the Sadducees were ignorant concerning the resurrection pursuant to which individuals are neither married nor given in marriage.

The failure of their "question and answer" strategy did not sway the Pharisees and Sadducees from their "killing Lazarus"

mindset. Eyewitnesses watched Jesus open
the eyes of a man who had been born blind.
The Pharisees commanded that the man be
brought before them to testify whether he
had indeed been born blind. The formerly
blind man testified that he had been blind
from his birth. The Pharisees refused to
believe his personal testimony and called
the man's parents. The parents were
reluctant to answer questions for fear of
being cast out of the synagogue. So, they
replied their son was, in fact, born blind but
was old enough to answer for himself. The
Pharisees again summoned the man in
question and told him to give the glory to
God because Jesus was a sinner for healing
on the Sabbath day. The man replied that

never before had anyone opened the eyes of one born blind and that Jesus must have been sent by God. The Pharisees then told him that he had been born in sin and did not qualify to teach them.

On another occasion, the Pharisees told Jesus that he cast out demons by the father of demons (called Satan). Jesus answered that a house divided against itself can not stand, and if Satan cast out Satan, his kingdom would come to an end.

Jesus shed further light upon the "killing Lazarus" mentality by talking about a certain rich man and a beggar. The beggar was a nuisance and lay at the rich man's gate covered with body sores. He wanted only the crumbs from the rich man's table. Finally, the beggar died and was carried by the angels into paradise. The rich man also died and looked across the great gulf between paradise and hell. He saw the beggar being comforted by Abraham. He begged Abraham to send the beggar to sprinkle some water on his tongue to diminish his suffering in the flames of hell. Abraham answered that it was not possible to cross the gulf between hell and paradise. The rich man then begged Abraham to send the beggar to his father's house to warn his brothers about the reality of hell. Abraham replied that his brothers should pay attention to Moses and the prophets. The rich man insisted that if someone from the dead warned them, then they would surely repent. Abraham assured him that if his brothers did not believe Moses and the prophets they would not be persuaded by one risen from the dead.

During his earthly ministry, Jesus made astounding statements. He claimed to be God in the flesh. He said he came into the world not to judge anyone but rather to make it possible to save mankind from eternal punishment for violence and depravity. Jesus said he would offer up his own body and blood to satisfy the penalty imposed by God's law. He said he would willingly lay down his life and then take it up again. He said man could not take his life but that he must lay it down of himself. In accordance with his own timing, Jesus allowed himself to be taken, abused, beaten, mocked, and crucified; and on the third day he rose from the dead according to his promise.

He was seen after his resurrection by his chosen apostles and several of the women who wept at his tomb. He spent forty days openly teaching his apostles following his resurrection. He also appeared before a gathering of more than five hundred individuals. He said that all who accept him as the sacrifice for their personal sins would be forgiven and would become the children of God. He further said that whenever humans accept him as their personal savior they will be given a new spiritual birth (being born again) and will love and reverence God rather than be at enmity with him.

Those directly responsible for killing Jesus claimed that his disciples secretly removed his body from his tomb and somehow disposed of it; that Jesus was just a common criminal; and that his followers were also criminals. They totally rejected Jesus as their prophesied Messiah. Their claim that some disciples hid the body of Jesus is entirely self-serving. The vehement murderers had obtained from Pontius Pilate a veteran Roman guard to watch over the body and the tomb was sealed with wax by Pilate's authority. The penalty for any Roman guard sleeping on duty was execution and breaking the seal was also a capital offense. Furthermore, more than five hundred eyewitnesses testified to seeing Jesus after his resurrection from the dead.

For more than two thousand years, the enemies of Jesus Christ have tried in vain to find one shred of evidence to support their refusal to accept him as God's sacrifice for their personal sins. They are willing to wade through the blood of God in the flesh in order to follow their spiritual father (Satan) into hell. They prefer the "prince of darkness" over the "prince of peace." Those entering hell in their earthly flesh will be there because they insisted upon their right to be there. They refused the free gift of redemption through the "Lamb of God" -- Jesus Christ. They nurtured the "killing Lazarus" mindset in spite of all the evidence to the contrary. Their eternal whining will go unheard.

The direct, eyewitness evidence that Jesus is precisely who he claimed to be is so overwhelming that every individual exposed to such evidence must either accept Jesus as God in the flesh or adopt the "killing Lazarus" mindset. Thus far, we have only considered direct, eyewitness evidence. There is also substantial circumstantial evidence supporting the eyewitness testimony. Recognized historians refer to various aspects of Jesus' earthly life and ministry. The most recognized and accepted historian referring to Jesus during the first century AD is Flavius Josephus who lived from AD 37-97. He was a Pharisee and served as a historian for the Roman Empire. He was born into a priestly Hebrew family and was not a follower of Jesus. Josephus wrote of Jesus that he was a wise teacher and a worker of many miracles; that his followers included both Jews and Gentiles; that Pilate, swayed by Hebrew high officials, condemned him to the cross; that those who loved him did not forsake him; and that his followers became known as "Christians."

Perhaps the most compelling circumstantial evidence is the sheer volume of Biblical prophecies foretelling the life, ministry, suffering, death, and resurrection of Jesus Christ. These prophecies are very specific and quite detailed and cannot refer to anyone other than Jesus -- God's Messiah to save humanity.

In 487 BC, it was prophesied that Messiah would be hailed as king upon his entry into Jerusalem as a humble servant mounted upon the foal of an ass. During the eleventh century BC, it was prophesied that the betrayer of Messiah would be a close friend who ate of Messiah's bread; that Messiah would be reviled, laughed at, tortured and put to death by his own kinsmen; that Messiah's hands and feet would be pierced; that Messiah would be given gall and vinegar during his suffering; and that his executioners would part his garments and cast lots upon his vesture.

In 712 BC, it was prophesied that Messiah would be scourged and his beard would be plucked out; that Messiah would be beaten and abused to the extent that he would not appear to be human; that Messiah would not open his mouth to complain while being oppressed and afflicted; and that Messiah would be executed in the company of criminals and be buried among the wealthy. In 487 BC, it was prophesied that Messiah would be betrayed for thirty pieces of silver and that the silver would later be cast down in the Hebrew temple and used to buy a field in which to bury strangers. These are not prophecies spoken in vague generalities but rather in such specific detail as to preclude every man who ever lived except Jesus Christ.

Humans have put forth only two explanations for the existence of our universe and all energy, matter, life, and motion which are perceived through the senses of vision, hearing, touching,

smelling and tasting. The oldest explanation dates back to the beginning of
recorded human history upon the planet,
Earth -- everything in the universe was
created by some supreme being or beings
(referred to as God or gods). First came the
belief in one God which degenerated over a
few centuries into belief in many gods to be
compatible with human lust, vanity and
perpetual wickedness (Sun God, Moon
God, Goddess of Fertility, Gods of War,
Gods of the Underworld, Love Goddess,
etc.). Idols depicting various invented gods
were carved from precious metals, wood,
stone, or clay. Idol worship generally
encompassed chanted prayers, human and
animal sacrifices, gluttony, sexual orgies,

self mutilation, and infanticide. The belief in either one God or plural gods has been perpetuated by the overwhelming majority of all humans upon Earth.

During the second half of the nineteenth century AD, a few individuals having the "killing Lazarus" mindset came up with the only other explanation as to how the universe and everything therein came into being -- the result of random, chaotic events wherein all energy, matter and motion popped out of nothingness and formed itself by random chance into the elements which concentrated all energy and matter into an extremely heavy walnut sized mass which exploded into galaxies, suns, moons, stars, planets, comets, gases, and all interstellar debris.

Then, they (Evolutionists) assumed that over billions of years, a piece of interstellar debris was formed by gravitational forces into the planet, Earth. About the same eon of time, a small, insignificant star attracted to itself Earth and the other eight planets within Earth's solar system and established the solar system's orbit around a Galaxy named "Milky Way." Earth's sun, being the largest object within Earth's solar system, determined the orbital period around the Milky Way. During additional eons of time, Earth's surface cooled while being bombarded by chunks of ice covered interstellar debris. Upon impact, the ice melted and formed Earth's oceans, rivers, lakes and other bodies of water. Earth's primitive oceans eroded rocks, minerals, and other compounds formed by the elements into a primordial soup which was bathed with light and energy from the sun.

Thereafter, within the primordial soup, perhaps triggered by a lightening strike, amino acids randomly clumped together and accidentally formed the first single-celled living organism from which over additional eons of time evolved every other life form including humans -- bacteria into aquatic creatures, aquatic creatures into amphibians, amphibians into mammals and reptiles, reptiles into birds, mammals into apes, and apes into humans.

This process of evolution perpetuated itself through random chance mutations wherein the weaker life forms became extinct and the stronger, better adapted life forms evolved from simplicity into complexity resulting in the most complex life forms existing today including homo sapiens. Therefore, God never did nor never will exist. Life on Earth is purely accidental and perpetuated by the stronger, better adapted life forms exploiting the weaker, less adaptable life forms and forcing these inferior creatures into extinction. Since there is no life after physical death, humans should copulate freely, eat, drink and be merry.

Evolution is today being taught in America's public schools as established fact supported by literally "mountains of evidence." Such proclamations by supporters of false science, false assumptions and false history are invariably accepted by the atheists controlling America's public schools. There is, however, not a single proven scientific fact to support the so-called "theory of evolution."

"When I consider the heavens, the work of Thy fingers, the moon and stars, which Thou hast ordained; what is man that Thou art mindful of him?" (Psalms 8:3,

1045 BC)

"For the invisible things of Him from the creation of the world are clearly seen, being under-stood by the things that are made, even His eternal power and Godhead; so that they are without excuse." (Romans 1:20, 60 AD)

".....we understand that the worlds were framed by the word of God, so that things which are seen were not made of things which do appear." (Hebrews 11:3, 64 AD)

Where did God come from? This is a reasonable and profound question which cannot be fathomed by humans outside the realm of spiritual beings who exist in unseen and unknowable dimensions. Human life does not perish with the death of the physical body because life and the body are separate entities. The physical body is composed of the chemical elements and life is not. At the death of the physical body, human life becomes part of the spiritual realm which is not comprehensible to the human spirit prior to separation from the physical body.

The entire physical history of the universe and Planet Earth is but one heartbeat within the timelessness of eternity such that the human mind cannot comprehend what lies beyond the pale of our brief mortality. We are simply incapable of rationalizing God or comprehending anything about His nature and power other than what Jesus Christ (God in human form) revealed to humanity during His earthly ministry and what is written down by divinely inspired authors and contained in the Holy Scriptures. Those who by human wisdom seek to search out and/or deny God are going to spend their eternal future in a very unpleasant habitation. Thus, it is written:

"Canst thou by searching find out God? Canst thou find out the Almighty unto perfection? It is as high as heaven; what canst thou do? Deeper than hell; what canst thou know?" (Job 11:7-8, 1520 BC)

"He hath made everything beautiful in His time: also He hath set the world in their heart, so that no man can find out the work that God maketh from the beginning to the end." (Ecclesiastes 3:11, 977 BC)

"O the depth of the riches both of the wisdom and knowledge of God! How unsearchable are His judgments, and His ways past finding out." (Romans 11:33, 60 AD)

"For we are but of yesterday, and know nothing, because our days upon earth are a shadow." (Job 8:9, 1520 BC)

"For a thousand years in Thy sight are but as yesterday when it is past, and as a watch in the night." (Psalms 90:4, 1015 BC)

"In the beginning was the Word, and the Word was with God, and the Word was

God. The same was in the beginning with God. All things were made by Him: and without Him was not any thing made that was made. In Him was life; and the life was the light of men. And the light shineth in darkness; and the darkness comprehended it not." (John 1:5, 30 AD)

"But, beloved, be not ignorant of this one thing, that one day is with the Lord as a thousand years, and a thousand years as one day.....But the day of the Lord will come as a thief in the night; in the which the heavens shall pass away with a great noise, and the elements shall melt with fervent heat, the earth also and the works that are therein shall be burned up." (II Peter 3:8 &10, 66 AD)

"And I saw a new heaven and a new earth: for the first heaven and the first earth were passed away; and there was no more sea......And He that sat upon the throne said, Behold I make all things new. And He said unto me, Write: for these words are true and faithful." (Revelation 21:1 -5, 96 AD)

"For My thoughts are not your thoughts, neither are your ways My ways, saith the Lord. For as the heavens are higher than the earth, so are My ways higher than your ways, and My thoughts than your thoughts." (Isaiah 55:8-9, 712 BC)

It is not a fruitful use of one's limited earthly existence to continue to try and shed light upon the rough pathway of time so that those insisting there is no God have adequate warning to forsake the darkness of human vanity; as it is written:

"Give not that which is holy unto the dogs, neither cast ye your pearls before swine, lest they trample them under foot, and turn again and rend you." (Matthew 6:6, 31 AD)

"Enter ye in at the strait gate: for wide is the gate, and broad is the way, that leadeth to destruction, and many there be that go in thereat: Because strait is the gate, and narrow is the way, which leadeth unto life, and few there be that find it." (Matthew 7: 13-14, 31 AD)

"And whosoever shall not receive

you, nor hear your words, when you depart out of that house or city, shake off the dust of your feet. Verily I say unto you, It shall be more tolerable for the land of Sodom and Gomorrah in the day of judgment than for that city." (Matthew 10:14-15, 31 AD)

Thus, it is highly unlikely that infidels, atheists, evolutionists and agnostics will be persuaded that their life in eternity will be rather bleak, hopeless and painful. Life is more full, fruitful and joyful when filled with love, charity, long suffering, forgiveness and empathy for those who know there is a God and who accept Jesus Christ as both the Incarnate Word and their personal sacrificial lamb providing redemption back to God, their

Father, from a state of sin and hopelessness.

Jesus, through His willing sacrificial death to redeem humanity made it possible for every human to come personally in prayer before God's throne of grace and mercy. Prior to His actual crucifixion only the Jewish High Priest could enter "the Holy place before the "mercy seat." A heavy curtain served as a veil separating the mercy seat from the rest of the temple interior. When Jesus died on His cross, this veil was torn down the middle indicating the way is now open to "whosoever will come freely believing."

Chapter five
Eternal love

Every human possesses a physical body composed of the primordial elements and is a complete entity in itself. This is an indisputable, scientific fact. **Common sense and simple logic** dictate that the life force within each individual human is most definitely not composed of the physical elements and therefore is not a part of nor dependent upon the existence of the physical human body. Each human life force is a spirit joined with an eternal soul. Dr. C. I. Scofield who created the Scofield Reference Bible best described the human trinity:

"Because man is "spirit" he is capable of God-consciousness and communion with God; because he is "soul" he has self-consciousness; because he is "body" he has, through his senses, world consciousness."

The spirit is the actual life force. The soul is the seat of emotions and self-will. The spirit and soul are immortal whereas the body is physical, mortal, and begins to die from birth. Every human spirit knows that the physical body it inhabits has an appointment with death because its body can see, touch and smell other human bodies returning to the elements from which they originated. However, each human spirit being a free moral agent can deny that a Creator exists.

Jesus Christ is the focus of the Holy Bible from the creation of man to the renovation and cleansing of Earth. Old Testament Scriptures point forward to His sacrificial death and new Testament Scriptures look back at His sacrificial death and subsequent resurrection. The Old Testament animal sacrifices offered up in accordance with the law of Moses were a prophetic rendering of the eternal sacrifice which God offered up of Himself through Jesus Christ in the fullness of time.

All redeemed humans from the creation of man to the end of measured time were and are forgiven by mentally accepting the sacrifice which God provided to pay the penalty for their sins. All the sacrifices offered under the law of Moses simply foreshadowed the coming of Jesus Christ. We are redeemed today by believing the record God gave us of the ministry, sacrificial death and subsequent resurrection of His Son, Jesus Christ (the incarnation of Almighty God)..

Old Testament people were redeemed back to God by believing in Him, trusting in His word delivered by the prophets, and by offering up blood sacrifices which were a substitute for the future sacrificial death and subsequent resurrection of God, the Son in the person of Jesus Christ. Not only is the shed blood of Jesus Christ the exclusive price paid to redeem humanity, the price of redemption was fully determin-ed before the creation of mankind:

"For as much as you know that you were not redeemed with corruptible things, as silver and gold, from your vain conversation received by tradition from your fathers; but with the precious blood of Christ, as of a lamb without blemish and without spot: who was foreordained before the foundation of the world, but was manifest in these last times for you. Who by Him do believe in God, that raised Him up from the dead, and gave Him glory; that your faith and hope might be in God." (1st Peter 1:18-21, 66 AD)

God, being an all powerful, all

knowing and creative spirit, comprehends the past, present and future simultaneously. Thus, God was not surprised by human disobedience (sin) in the beginning of measured time. God had foreordained the price of human redemption and the Son was willing to pay the price.

The only part humans play in the divine plan of redemption (salvation) is the belief in and acceptance of the sacrifice which God provided of Himself in the person of Jesus Christ. Therefore, during unmeasured eternity, God will have exactly what He envisioned when He created humans.....living beings in His Own Image who love, reverence and fellowship with Him because they choose to do so.

Paul, the Apostle explains the logic and justice of human redemption through our sacrificial lamb (Jesus Christ) at Romans 5:12-21, 60 AD:

"Wherefore, as by one man sin entered into the world, and death by sin; and so death passed upon all men for that all have sinned: For until the law (*law of Moses*) sin was in the world: but sin is not imputed when there is no law. Nevertheless death reigned from Adam to Moses, even over them that had not sinned after the similitude of Adam's transgression, who is the figure of Him that was to come. But not as the offense, so also is the free gift. For if through the offense of one many be dead, much more the grace of God, and

the gift by grace, which is by one man,
Jesus Christ, has abounded unto many.
And not as it was by one that sinned, so is
the gift: for the judgment was by one to
condemnation, but the free gift is of many
offenses into justification. For if by one
man's offense death reigned by one; much
more they which receive abundance of
grace and of the gift of righteousness shall
reign in life by one, Jesus Christ. Therefore
as by the offense of one judgment came
upon all men to condemnation; even so by
the righteousness of one the free gift came
upon all men unto justification of life. For
as by one man's disobedience many were
made sinners, so by the obedience of one
shall many be made righteous. Moreover,

the law entered, that the offense might abound *(humanity aware of their sinning against God);* but where sin abounded, grace did much more abound. That as sin has reigned onto death, even so might grace reign through righteousness unto eternal life by Jesus Christ our Lord." (Romans, 5:12-21, 60 AD)

In this single passage is summed up the exact purpose and content of God's word penned down in the Holy Bible. There is no excuse for any human to reject God's free gift of eternal life in His kingdom because God requires nothing but individual acceptance of the gift of redemption and subsequent eternal life in fellowship with the Godhead as opposed to an eternal existence in "the lake of fire prepared for Satan and his followers."

Every human since Adam has followed either God or Satan through the exercise of his/her free will. The good works that Christians strive for are not for the purpose of redemption but out of love, respect, adoration, and the fervent desire to be pleasant in the sight of our Heavenly Father. All those who join Satan in his eternal punishment will be doing so by individual choice.

Redemption is totally free whereas heavenly rewards are given out for overcoming the lust of the eyes, the lust of the flesh and the pride of life (often described as "faithful stewardship").

Jesus Christ is the central theme of the Holy Bible. The first thirty-nine books making up the Old Testament look forward to the human birth of Jesus and the twenty-seven books comprising the New Testament look back at the cross where Jesus offered up Himself as God's sacrificial lamb to redeem fallen humanity back to Himself.

Thus, the Bible is not a scientific rendition but rather the written documentation of God's eternal plan of human salvation. However, the Bible reveals what humans now refer to as "laws of physics" more than two thousand years before scientists discovered such physical relationships. The birth, ministry, sacrificial death and resurrection of Jesus Christ was pictured in the Bible within numerous passages written down between 1500 BC and 400 BC spanning the historical window between the Hebrew slavery in Egypt and the prophecies written by Malachi which closed the Old Testament cannon.

The Old Testament contains the divine covenant of God's holy laws and the New Testament records the divine covenant of grace and mercy which was a mystery hidden within the covenant of law. God's **law** was given to Moses but **grace and truth** was revealed through Jesus Christ.

Over a time window of sixteen centuries God, through the Old Testament prophets, brushed aside the curtain of time to picture Jesus Christ in astounding details:

"Therefore the Lord Himself shall give you a sign; Behold, a virgin shall conceive and bear a son, and shall call His name "Immanuel" (meaning God with us). [Isaiah 7:14 KJV; 742 BC]

The birth prophecy continues in Isaiah 9:6-7: "For unto us a Child is born, unto us a Son is given: and the government shall be upon His shoulder: and His name shall be called Wonderful, Counselor, The Mighty God, The Everlasting Father, The Prince of Peace. Of the increase of His government and peace there shall be no end, upon the throne of David, and upon His kingdom, to order it, and to establish it with judgment and with justice from henceforth even for ever."

Of course, Jesus being born to serve as God's Sacrificial Lamb was familiar with Isaiah's prophecies as were the scribes, Pharisees, high priest, chief priests and other Jewish religious rulers in the days of King Herod the Great and the Roman occupation of the Holy Land.

The primary difference between what Jesus knew and what those plotting to murder Him knew is that Jesus was fully aware He would be offered up on a Roman cross to atone for the sins of humanity before returning to Earth as King of Kings and Lord of Lords.

The Jews were steeped in traditional thought and customs and were expecting their Messiah to break the Roman yoke and establish Jerusalem as the center of world government. Even so, the prophecies spoken and written by Isaiah did not lend themselves to manipulation by Jesus as an impostor or by those Jews who plotted His death. God showed the prophet, Isaiah, Jesus Christ being hailed as "King of the Jews" and the Messiah, and then gave him a glimpse of Jesus enduring unmerciful torture, abuse, and mockery:

"Behold, My servant shall deal prudently, He shall be exalted and extolled, and be very high. As many were astonished at Thee; His visage was so marred more than any man, and His form more than the sons of men." (Isaiah 52:13-14; 712 BC)

"I gave My back to the smiters, and My cheeks to them that plucked off the hair: I hid not my face from shame and spitting." (Isaiah 50:6; 712 BC)

Then in Isaiah, Chapter 53, the prophet draws a picture that is forever engraved upon the author's heart:

"Who hath believed our report? And to whom is the arm of the Lord revealed? For He shall grow up before Him as a tender plant, and as a root out of dry ground: He hath no form nor comeliness; and when we see Him, there is no beauty that we should desire Him. He is despised and rejected of men; a man of sorrows, and acquainted with grief: and we hid as it were our faces from Him; He was despised, and

we esteemed Him not. Surely He hath
borne our griefs, and carried our sorrows:
yet we did esteem Him stricken, smitten of
God, and afflicted. But He was wounded
for our transgressions, He was bruised for
our iniquities: the chastisement of our
peace was upon Him; and with His stripes
we are healed. All we like sheep have gone
astray; we have turned every one to his own
way; and the Lord has laid upon Him the
iniquity of us all. He was oppressed, and
He was afflicted, yet He opened not His
mouth: He is brought as a lamb to the
slaughter, and a sheep before her shearers is
dumb, so He opened not His mouth. He
was taken from prison and from judgment:
and who shall declare His generation? For

He was cut off out of the land of the living:
for the transgressions of my people was He
stricken. And He made His grave with the
wicked, and with the rich in His death;
because He had done no violence, neither
was there any deceit in His mouth.

Yet it pleased the Lord to bruise Him; He
hath put him to grief: when Thou shall shalt
make His soul an offering for sin, He shall
see His seed, He shall prolong His days,
and the pleasure of the Lord shall prosper
in His hand. He shall see of the travail of
His soul, and shall be satisfied: by His
knowledge shall My righteous servant
justify many; for He shall bear their
iniquities. Therefore will I divide Him a
portion with the great, and He shall divide

the spoil with the strong: because He hath poured out His soul unto death: and He was numbered with the transgressors; and He bare the sin of many, and made intercession for the transgressors." (Isaiah 53:1-12; 712 BC)

In Isaiah, Chapter 61:1-3; 698 BC, we are given a preview of the earthly ministry of Jesus as He sets His face as flint toward His cross:

"The Spirit of the Lord God is upon Me; because the Lord hath anointed Me to preach good tidings to the meek; He hath sent Me to bind up the brokenhearted, to proclaim liberty to the captives, and the opening of the prison to them that are bound; to proclaim the acceptable year of the Lord, and the day of vengeance of our God; to comfort all that mourn; to appoint to them that mourn in Zion, to give unto them beauty for ashes, the oil of joy for mourning, the garment of praise for the spirit of heaviness; that they might be called trees of righteousness, the planting of the Lord, that He might be glorified."

The prophet Zechariah predicted the precise price of the betrayal of Jesus by Judas:

"And I said unto them, if ye think good, give me my price; and if not, forbear. So they weighed for My price thirty pieces of silver. And the Lord said unto me, Cast it unto the potter: a goodly price that I was prised at of them. And I took the thirty pieces of silver, and cast them to the potter in the house of the Lord." (Zechariah, 11:12-13; 487 B.C.).

520 years later, the remorse of Judas is recorded:

"Then Judas, which had betrayed Him, when he saw that He was condemned, repented himself, and brought again the thirty pieces of silver to the chief priests and elders, saying, I have sinned in that I have betrayed innocent blood. And they said, What is that to us? See thou to that. And he cast down the pieces of silver in the temple, and departed, and went and hanged himself. And the chief priests took the silver pieces, and said, it is not lawful for to put them into the treasury, because it is the price of blood. And they took counsel, and bought with them the potter's field, to bury strangers in." (Matthew, Chapter 12:3-7; 33 AD)

The prophet Micah prophesied the precise hamlet where Jesus would be born: "But thou, Bethlehem Ephratah, though thou be little among the thousands of Judah, yet out of thee shall He come forth unto Me that is to be ruler in Israel; whose goings forth have been from old, from everlasting." (Micah, Chapter 5:2, 710 B.C.)

More than 1,000 years before the birth of Jesus, the Spirit of God Almighty stirred King David's soul and he penned down in Psalms 22:

"My God, My God, why hast thou forsaken Me? Why art Thou so far from helping Me, and from the words of My roaring? ……….But I am a worm, and no man; a reproach of men, and despised of the people. All that see Me laugh Me to scorn: they shoot out the lip, they shake their head, saying, He trusted on the Lord that He would deliver Him: let Him deliver Him, seeing He delighted in Him………. They gaped upon Me with their mouths, as a ravening and a roaring lion. I am poured out like water, and all My bones are out of joint: My heart is like wax; it is melted in the midst of My bowels. My strength is dried up like a potsherd; and My tongue cleaveth to My jaws; and Thou hadst brought Me unto the dust of death. For dogs have compassed Me: the assembly of the wicked have enclosed Me: they pierced My hands and My feet. I may tell all My bones: they look and stare upon Me. They part My garments among them, and cast lots upon My vesture." (Portions of Psalms 22; 1017 BC)

In Psalms 69; 1017 BC, King David again writes down what God lays upon his heart concerning the suffering of Jesus as despised and rejected:

"They that hate Me without a cause are more than the hairs of Mine head: they that would destroy Me, being Mine enemies wrongfully, are mighty; Then I restored that which I took not away......

Reproach hath broken My heart; and I am full of heaviness: and I looked for some to take pity, but there were none; and for comforters, but I found none. They gave Me also gall for My meat; and in My thirst they gave Me vinegar to drink."

These verses in Psalms 69 paint an exact portrait of Jesus upon His cross reuniting humanity with God while being ridiculed, mocked, jeered at and given vinegar mixed with gall to drink.

The Holy Bible describes the fall of the angel Lucifer who we now refer to as "Satan."

"How have you fallen from heaven, O Lucifer, son of the morning!.....For you have said in your heart: I will ascend into

heaven, I will exalt my throne above the
stars of God: I will sit also upon the mount
of the congregation, in the sides of the
north; I will ascend above the heights of the
clouds; I will be like the Most High.
(Isaiah, 14:12-14: 712 BC)

".......Thus saith the Lord God: You
seal up the sum, full of wisdom, and perfect
in beauty. You have been in Eden the
garden of God; every precious stone was
your covering, the sardius, topaz, and
diamond, the beryl, the onyx, and the
jasper, the sapphire, the emerald, and the
carbuncle, and gold: the workmanship of
your tabrets and of your pipes was prepared
in you in the day that you were created.
You are the anointed cherub that covers;

and I have set you so; you were upon the holy mountain of God; you have walked up and down in the midst of the stones of fire. You were perfect in your ways from the day that you were created, until iniquity was found in you." (Ezekiel, 28:12-15; 588 BC)

"And there was war in heaven: Michael and his angels fought against the dragon; and the dragon fought and his angels, and prevailed not; neither was their place found any more in heaven. And the great dragon was cast out, that old serpent called the Devil, and Satan, which deceives the whole world: he was cast out into the earth, and his angels were cast out with him." (Revelation, 12:7-9; 96 AD)

The composite of these passages implicates Earth in the "fall of Lucifer," (also called "Satan," "Devil." and "Dragon"). There was an absence of energy lighting the planet, and there was no division of liquids and gases on Earth's surface. Whatever living creatures inhabited Earth at that time perished as well as plant life. However, plant seed remained dormant within Earth. The physical shape of Earth reflected chaos such that it appeared "without form and void," and completely covered with water.

The intense, eternal hatred Lucifer exhibits toward mankind is understandable in the light of these related Scriptures. The dominion of Earth was given to humans and Lucifer is being allowed to sift and test human obedience and reverence for God. When the first created humans rebelled against God in their garden paradise, they were cast out under sentence of physical death into a hostile environment ruled by Satan to whom they had yielded up the dominion of Earth. They were banished from God's presence appointed to physical death but carried with them the "breath of God" (their immortal life force).

Consider the uniqueness of the first male and female humans (Adam and Eve). Their physical bodies were created from the primordial elements by a supreme life bequeathing spirit. Both their bodies and their spirits were immortal. They were created to live forever free of disease, aging and physical death within a perfect paradise over which they were given complete dominion. They were the undisputed masters of Planet Earth and could beget sons and daughters in their own image who would also be immortal. They were created in God's image after His likeness and they had everything mortal, fallen humans can only dream of today.

"And God said, 'Let us make man in our image, after our likeness: and let them have dominion over the fish of the sea, and over the fowl of the air, and over the cattle, and over all the earth, and over every creeping thing that creepeth upon the earth.' So God created man in His own image, in the image of God created He him; male and female created he them. And God blessed them, and God said unto them, 'Be fruitful and multiply, and replenish the earth, and subdue it: and have dominion over the fish of the sea, and over the fowl of the air, and over every living thing that moveth upon the earth.'-" (Genesis 1:26-28, 4004 BC)

The only thing that could affect their eternal life was to exercise their free will and do the one and only thing God commanded them not to do. It was the way God chose to test whether Adam and Eve would be lifted up with pride and arrogance like Lucifer.

The law personally given by God to humans through Moses was not in effect until 1491 B.C. Where there is no law, there is no sin; for sin, **by definition,** is the transgression of God's law. Thus, in the Garden of Eden, God's entire law was one simple commandment: "And the Lord God commanded the man saying, 'Of every tree of the garden thou mayest freely eat: But of the tree of the knowledge of good and evil, thou shalt not eat of it: for in the day that thou eatest thereof thou shalt surely die.'-" (Genesis, chapter 2, verses 16-17 4004 BC).

The forbidden tree was the only vehicle through which Adam and Eve could come to know the folly of sin and the loss of pure innocence. Moreover, they also knew the penalty for sin is death. Through sin against God they would become vulnerable to the living creatures opposing God (Lucifer and his angels). It is therefore most fitting that God called the forbidden tree the "tree of the knowledge of good and evil." The tree, itself, was of little significance. It was the act of willing disobedience coupled with the knowledge of the penalty that would "open their eyes" to the knowledge of good and evil and result in spiritual banishment from God's presence followed by physical death. Hence, it was no small thing, this eating of the forbidden fruit.

Why did Adam and Eve, knowing the penalty, willingly break God's single commandment given to them? First, Eve listened to Lucifer who is the father of lies: ".....Ye shall not surely die: For God doeth know that in the day ye eat thereof, then your eyes shall be opened, and ye shall be as gods, knowing good and evil." (Genesis, 1:4-5 , 4004+? BC)

Then, Eve allowed herself the twin pleasures of lust and pride. The fruit was pleasant to look at, it would probably be very tasty, and it would make her as wise as God. Besides, the serpent was more to be trusted than her creator. God had been holding out on her. Pride turned into arrogance. She reached out and fondled death. It felt just like a ripe, delicious fruit. Her breasts swelled with rebellion and pride as she hurried to find Adam to give him a taste. For the very first time, Eve became aware of the force of human lust and her own nakedness. Should Adam be less than enthusiastic about risking death, she could probably coax him into joining her.

Adam and Eve became spiritually fallen that very day and were cast out of the garden of God. They were banished from God's presence to labor for their food and to conceive children in their own image of lust, pride and arrogance. They were now appointed to physical death and their children would be under the same death sentence. Immortal had become mortal, innocence had become lust; eternal physical life had been left behind in the garden of God.

Mortal living beings cannot pass on eternal physical existence. Adam and Eve had sealed the fate of their children and all future generations. All would be born in a state of sin and mortality. All human flesh was forever barred from God's eternal presence.

But hope still existed for that breath of God which imparted, through Adam and Eve, an eternal living soul to every human for all eternity. Even when casting Adam and Eve from the garden into a hostile world where sin would abound, God knew He would incarnate Himself and offer up His own body and blood to redeem human souls. The very essence of God is the pure divine love of our Heavenly Father for His human children. The incarnation of God into human flesh in the person of Jesus Christ is the central truth of the Bible and also is the acid test for Scriptural infalli-bility. If Jesus Christ is, in fact, who He claims to be, then the search for truth is over. Jesus not only said He is the son of God. He said he is God:

"I and my Father are one." (John, chapter 10, verse 30) ".....he that hath seen me hath seen the Father...." (John 14:9, 33 AD)

"In the beginning was the Word, and the Word was with God, and the Word was God" (John, chapter 1, verse 1). ".....and the Word was made flesh and dwelt among us,......" (John 1:1-14, 26 AD)

"For God so loved the world, that He gave His only begotten Son, that whosoever believes in Him should not perish, but have everlasting life. For God sent not His Son into the world to condemn the world; but that the world through Him might be saved." (John 3:16-17, 30 AD)

Chapter six
All things new

Being prone to iniquity is not due to defective genetics nor any other biological factor. It is due to the willing choice of Satan as spiritual father when making the free-will choice between God and Satan.

All humans are born into a cursed planet ruled by Satan to whom Adam and Eve voluntarily surrendered dominion of Earth. God will, in accordance with His ownership of the entire universe, create a new heaven and a new Earth for His spiritual children. In the meanwhile, God allows Satan to force the free-will choice

by all humans as to their spiritual father.
Satan's spiritual children will follow him
into the "lake of fire" prepared for him and
his voluntary followers who must trample
the blood of Jesus Christ under foot on their
way to eternal banishment in outer
darkness.

Within the human trinity of body,
soul and spirit, an internal spiritual war is
raging between the Heavenly Trinity of
God, Jesus Christ and the Holy Spirit
versus Satan, fallen angels and demonic
spirits led by Satan while he exercises
dominion of Earth which he stole from
Adam and Eve through deception and
accusation. The demonic spirits are some of
the fallen angels who exercised their free

will to join Satan's attempt to usurp God's dominion of the entire universe. The rest of the fallen angels are imprisoned in outer darkness awaiting judgment by God. The Apostle Paul, who wrote half of the books in the Biblical New Testament, expressed the battle for human souls in the Epistle to the Romans:

"For that which I do I allow not; but what I hate I do. If then I do that which I would not, I consent unto the law that it is good. Now then it is no more I that do it, but sin that dwelleth in me. For I know that in me (that is, in my flesh) dwelleth no good thing: for to will is present with me; but how to perform that which is good I find not. For the good that I would I do not:

but the evil which I would not, that I do.
Now if I do that I would not, it is no more I
that do it, but sin that dwelleth in me. I find
than a law, that, when I would do good, evil
is present with me. For I delight in the law
of God after the inward man: But I see
another law in my members, warring
against the law of my mind, and bringing
me into captivity to the law of sin which is
in my members. O Wretched man that I
am! Who shall deliver me from the body of
this death? I thank God through Jesus
Christ, our Lord. So then with the mind I
myself serve the law of God; but with the
flesh the law of sin. There is therefore now
no condemnation to them which are in
Christ Jesus, who walk not after the flesh,

but after the Spirit. " (Romans 7:15-25; 8:1, 60 AD)

How does the Bible teach that a believer in Jesus Christ should view the law of Moses? A believer should view the law given to Moses by God as having fulfilled the purpose for which it was given..... to make individuals aware of the need of every human for a sacrificial lamb for redemption back to God and forgiveness of personal sins past, present and future. Where there is no law, there is no sin. A believer is brought out from under the law and redeemed by grace through faith in Jesus Christ. The law simply does not apply to believers for the purpose of judgment but rather as a guide as to how a believer can satisfy the desire to please God complete with the full knowledge that no human being in a mortal, physical body can keep the perfect law of God. Otherwise Jesus Christ died in vain. Nevertheless, right believing (in Jesus Christ) will lead to the undeniable desire to love, honor, worship and please God

Consequently, it is no longer the "sin question" but rather the "Son question:" "What think ye of Christ?" Did you believe in him and accept him as your personal sacrificial lamb? If not, you remain under the law given by God to Moses and you will follow your spiritual father into his eternal habitat of banishment and torment.

However, you will be there in

accordance with your free-will choice to reject Jesus Christ as your "sacrificial lamb" in full payment of your sin debt pertaining to God and your fellow humans. The willing sacrificial death of Jesus Christ made it possible for God to dispense righteous judgment of sin and divine mercy simultaneously.

The divine love of God toward humanity is beyond human comprehension. God's plan of redemption and eternal salvation for humans required redemption by a near kinsman who was willing to pay the full penalty for human rebellion and disobedience. God, Himself, was the only near kinsman who was both capable and willing to pay such a price.

The transgression of Adam and Eve was a monumental offense. They had lived in God's personal presence. They had walked and talked with God. They had intimate knowledge of God as creator, Heavenly Father and friend. They knew the penalty for rebellion against God's single commandment was spiritual banishment coupled with physical death. They knew spiritual banishment would be immediate and that their physical bodies would be appointed to eternal death within their mortal life span.

They knew all this and yet they defied God and did the only single thing God said not to do under penalty of death. The price to redeem them back to God was astronomical; otherwise divine justice would be meaningless. God, through Jesus Christ, took upon Himself human form and paid the price in full to redeem humans back to Himself. Thus, human salvation is a totally free gift lavished upon humanity by a loving, merciful and gracious Supreme Being who takes no pleasure in the death of the wicked.

During the course of human history, in each generation, individual humans have personally chosen to believe in random spontaneous generation of life forms or to believe in an all-powerful Creator.

Spontaneous generation of any life form within the universe has been proven to be scientifically as well as mathematically impossible. Nevertheless, those humans who reject, for whatever personal reason, the existence of a Creator existing outside of time and space, choose to believe that everything accidentally sprang out of nothingness without design, order or purpose even though they know that such a concept is absolutely impossible.

In order to cope with a personal belief in an impossible concept, many humans have decided that they are living in an imaginary universe and that nothing is real. Others have decided that some sort of Creator does exist but is unapproachable and thus unknowable.

Others have soothed their need to commune with a Creator by worshiping idols they create from clay, wood, stone and metal; or choose to worship the sun, moon, stars, etc. Some choose to worship themselves or lower life forms.

It is not uncommon for humans to claim that they have made no decision regarding the existence of a creator. There is no state of being void of a conscious decision.

Before any word can be spoken or any physical act undertaken, the relevant thought must first take shape in the brain. Therefore, humans think, say and act based on what they have chosen to believe. Thoughts, words and acts that are contrary to personal beliefs result in sorrow, grief, anxiety, depression, and sometimes, a

change in what the individual chooses to
believe.

A personal belief system which
incorporates an all-powerful Creator
existing outside of time and space is
usually referred to as a "religion." Over six
billion humans alive today have adopted a
specific religion. Most of the adherents of
any religion adopt whatever belief system
is taught to them as a child.

Once a religion is adopted, it is rare
that the individual turns to another religion.
It is more common for an individual to
discard a religious belief and turn to
atheism.

Approximately one third of the
world's population profess to be Christians

and believe in the deity of Jesus Christ. The sacred text for Christianity is the sixty-six divisions of the Holy Bible written by forty different authors between 1491 B.C. and 96 A.D.

The most astounding fact supporting the absolute veracity of the Holy Bible is that forty different humans over a period spanning nearly sixteen centuries wrote in total harmony with each other concerning the creation of the universe and all its life forms. They also described the same Creator existing outside of time and space and the eternal relationship between the Creator and humans.

In addition, Biblical prophets predicted hundreds of specifically detailed

events spanning twenty-four centuries
which came to pass exactly as predicted.
Biblical authors referred to the all-powerful
creator as Jehovah, Yahweh, Adonai, and I
AM THAT I AM.

The English translation of the Holy
Bible refers to the same single all-powerful
deity as Lord, Lord God, Lord of Hosts,
Creator, Almighty God, Jesus Christ, Lamb
of God, Son of God, Son of Man, Holy
Spirit, and Holy Ghost.

The Holy Bible declares repeatedly
that the single all-powerful deity reveals
Himself to humanity in both spirit and
human form as God, the Father; God, the
Son; and God, the Holy Spirit.

From the perspective of humans on

Planet Earth pertaining to the hierarchy of intelligent life forms, there exists only demons and angels between humans and the Triune Godhead. Humans have heard God's audible voice and seen Him in both angelic (Angel of the Lord) and human form (Jesus Christ).

Many "lip service" Christians today reject the concept of "hell" and the "lake of fire." They might as well use their Bibles for toilet paper. The Scriptures are very clear on the subject:

"And many of them that sleep in the dust of the earth shall awake, some to everlasting life, and some to shame and everlasting contempt." (Daniel, chapter 12, verse 2, 534 BC)

"And death and hell were cast into the lake of fire. This is the second death." (Revelation, chapter 20, verse 14, 96 A.D.)

"But the fearful, and unbelieving, and the abominable, and murderers, and whore mongers, and sorcerers, and idolaters, and all liars, shall have their part in the lake which burneth with fire and brimstone: which is the second death." (Revelation, chapter 21, verse 8, 96 A.D.)

"And the smoke of their torment ascendeth up for ever and ever: and they have no rest day nor night......(Revelation, chapter 14, verse 11, 96 A.D.)

Hell is to the lake of fire as a local holding cell is to the state prison. Unbelievers are emptied out of hell to be formerly judged before being cast into the lake of fire. Hell is temporary confinement in a very unpleasant place; the lake of fire is eternal. Hell, after giving up its inhabitants, will lose its significance.

Where is hell? It really doesn't matter to those who believe in Jesus Christ as the divine sacrifice for their sins because hell will greet only those who trample His sacrifice under foot and insist on being held for final judgment. All who enter hell, and the lake of fire, will do so in spite of everything God has done to keep them out of both places.

The question is often asked: how can a loving God send anyone to hell? The answer is: people go to hell because they

want to spend eternity with their spiritual
father, Satan. There is no need for anyone
to go to hell other than by choice. Jesus
Christ paid for the sins of all humanity.
People go to hell for only one reason --
refusal to accept forgiveness for their
personal sins. Do not blame God for
winding up in the lake of fire; it is simply a
matter of choice.

There are several Scriptures which
indicate hell is located inside the planet.
The outer core of Earth is extremely hot
and full of molten rocks and metals. Thus,
the general environment is suitable. Hell is
also described as the bottomless pit. Earth,
being a sphere has no bottom. Hell is
further described as "beneath" with

reference to Earth's surface. Then, there is this Scripture:

"They, and all that appertained to them, went down alive into the pit, and the earth closed upon them: and they perished from among the congregation." (Numbers, chapter 16, verse 33; 1471 B.C.)

The lake of fire could be any flaming mass within the universe. There are literally billions of flaming masses, which we call stars, that would serve the purpose quite well.

Jesus, when He prayed to the Father, sometimes looked toward the heaven within which the throne of God is located: "These words spake Jesus, and lifted up His eyes to heaven, and said, Father, the hour is come; glorify Thy Son, that thy Son also may glorify Thee." (John, chapter 17, verse 1, 33 AD)

Just as hell is beneath, with reference to Earth's surface, heaven is above. From any point on Earth in either hemisphere, we can gaze into space, day or night. The vastness of space is beyond description except in terms of light years. One light year is a distance spanning approximately 5.87 trillion miles. Where, in relation to Earth, is the "heaven of God" located? Looking toward the north in space through a powerful telescope, we see what is not apparent to the naked eye.

There is a huge area in the north where there are no stars. This starless area encompasses millions of light years with respect to dimensions. Isn't that interesting.

It becomes extremely interesting when the following Scripture is considered: "How art thou fallen from heaven, O Lucifer, son of the morning! How art thou cut down to the ground, which didst weaken the nations! For thou hast said in thine heart, I will ascend into heaven, I will exalt my throne above the stars of God: I will sit also upon the mount of the congregation, in the sides of the north........" (Isaiah, chapter 14, verses 12-13, written 712 B.C.)

Notice that Lucifer desires to "ascend" which indicates an upward direction. He wants his throne "above the stars of God" indicating a place without stars. He also wants to sit upon the "mount of the congregation" which is in the "sides of the north."

The staggering distances within space are prohibitive to mortal beings who function in the three dimensions of space, distance and time. However, to immortal beings, such considerations are trivial. The "heaven" where God's throne is located could, of course, be a region totally unknown to humans. But, again, what does it matter? It is the content and not the location which merits our attention. Immortal, spiritual beings are not bound by physical considerations such as space, distance and time.

The opening of the *"last days prophetic time window"* was prophesied by Isaiah, Jeremiah, Ezekiel, and Daniel:

"Thus saith the Lord God, Behold, I will lift up Mine hand to the Gentiles, and set My standard to the people: and they shall bring thy sons in their arms, and thy daughters shall be carried upon their shoulders. And kings shall be thy nursing fathers, and their queens thy nursing mothers: they shall bow down to thee with their faces toward the earth, and lick up the dust of thy feet; and thou shalt know that I am the Lord: for they shall not be ashamed that wait for Me. Shall the prey be taken from the mighty, or the lawful captive delivered? But thus saith the Lord, Even the

captives of the mighty shall be taken away,
and the prey of the terrible delivered: for I
will contend with him that contendeth with
thee, and I will save thy children. And I
will feed them that oppress thee with their
own flesh; and they shall be drunken with
their own blood, as with sweet wine: and
all flesh shall know that I the Lord am thy
Savior and thy Redeemer, the Mighty One
of Jacob." (Isaiah 49:22-26 712 B.C. KJV)

"Rejoice ye with Jerusalem, and be
glad with her, all ye that love her: rejoice
for joy with her, all ye that mourn for her:
That ye may suck, and be satisfied with the
breasts of her consolations; that ye may
milk out, and be delighted with the
abundance of her glory. For thus saith the

Lord, Behold, I will extend peace to her like a river, and the glory of the Gentiles like a flowing stream: then shall ye suck, ye shall be borne upon her sides, and and be dandled upon her knees. As one whom his mother comforteth, so will I comfort you; and ye shall be comforted in Jerusalem." (Isaiah 66:10-13 698 B.C. KJV)

"Therefore, behold, the days come, saith the Lord, that they shall no more say, The Lord liveth, which brought up the Children of Israel out of the land of Egypt; But, the Lord liveth, which brought up and which led the seed of the house of Israel out of the north country, and from all countries whither I had driven them; and they shall dwell in their own land."

(Jeremiah 23:7-8 599 B.C. KJV)

"For thus saith the Lord, Sing with gladness for Jacob, and shout among the chief of the nations: publish ye, praise ye, and say, O Lord, save thy people, the remnant of Israel. Behold, I will bring them from the north country, and gather them from the coasts of the earth, and with them the blind and the lame, the woman with child and her that travaileth with child together: a great company shall return thither. They shall come with weeping , and with supplication will I lead them: I will cause them to walk by the rivers of waters in a straight way, wherein they shall not stumble: for I am a Father to Israel, and Ephraim is My firstborn. Hear the word of the Lord, O ye nations, and declare it in the isles afar off, and say, He that scattereth Israel will gather him, and keep him, as a shepherd doth his flock. For the Lord hath redeemed Jacob, and ransomed him from the hand of him that was stronger than he." (Jeremiah 31:7-11 606 B.C. KJV)

"Therefore say, Thus saith the Lord God; I will even gather you from the people, and assemble you out of the countries where ye have been scattered, and I will give you the land of Israel. And they shall come thither, and they shall take away all the detestable things thereof and all the abominations thereof from thence. And I will give them one heart, and I will put a new spirit within them; and I will take the stony heart out of their flesh, and will give them a heart of flesh: That they may walk in My statutes, and keep Mine ordinances, and do them: and they shall be My people, and I will be their God." (Ezekiel 11:17-20 594 B.C. KJV)

"For thus saith the Lord God; Behold, I even I will both search My sheep, and seek them out. As a shepherd seeketh out his flock in the day that he is among his sheep that are scattered; so will I seek out My sheep, and will deliver them out of all places where they have been scattered in the cloudy and dark day. And I will bring them out from the people, and gather them from the countries, and will bring them to their own land, and feed them upon the

mountains of Israel by the rivers, and in all the inhabited places of the country."
(Ezekiel 34:11-13 587 B.C. KJV)

"Therefore thus saith the Lord God; Now will I bring again the captivity of Jacob, and have mercy upon the whole house of Israel, and will be jealous for My holy name: After that they have borne their shame, and all their trespasses whereby they have trespassed against Me, when they dwelt safely in their land, and none made them afraid. When I have brought them again from the people, and gathered them out of their enemies' lands, and am sanctified in them in the sight of many nations; Then shall they know that I am the Lord their God, which caused them to be led into captivity among the heathen: but I have gathered them into their own land, and have left none of them any more there. Neither will I hide My face any more from them: for I have poured out My Spirit upon the house of Israel, saith the Lord God." (Ezekiel 39:25-29 587 B.C. KJV)

"And from the time that the daily sacrifices shall be taken away, and the abomination that maketh desolate set up, there shall be a thousand two hundred and ninety days. Blessed is he that waiteth, and cometh to the thousand three hundred and five and thirty days." (Daniel 12:11-12 534 B.C. KJV)

God did **not** give the prophets the elapsed time between May 14, 1948 (the exact date of the **rebirth of Israel** as a sovereign nation) and the war between the nations from which Israel will emerge as a formidable nuclear power forcing even the Anti-Christ to sign a peace treaty with them. Ezekiel and Zechariah prophesied concerning the war between the nations using descriptions of armies and military weapons that people in 587 B.C. (Ezekiel) and 487 B.C. (Zechariah) could relate to. Nouns and descriptive phrases such as hydrogen bombs, fighter jets, helicopters, tanks, artillery pieces, rockets, missiles, paratroopers, long range bomber and carrier aircraft would have meant nothing to people six to seven centuries before Christ. Neither would the prophets have known words to communicate 21st century A.D. warfare:

"And the word of the Lord came unto me, saying, Son of man, set thy face against Gog, the land of Magog, the chief prince of Meshech (*Moscow*) and Tubal (*Tobolsk*), and prophesy against him, And say, Thus saith the Lord God; Behold, I am against thee, O Gog, the chief prince of Meshech and Tubal: And I will turn thee back, and put hooks into thy jaws, and I will bring thee forth, and all thine army, horses and horsemen, all of them clothed with all sorts of armor, even a great company with bucklers and shields, and all of them handling swords: Persia, Ethiopia, and Libya with them: all of them with shield and helmet: Gomer, and all his bands; the house of Togarmah of the north quarters, and all his bands: and many people with thee. Be thou prepared, and prepare for thyself, thou, and all thy company that are assembled unto thee, and be thou a guard unto them. After many days thou shall be visited: in the latter years thou shalt come into the land that is brought back from the sword, and is gathered out of many people, against the mountains of Israel, which have always been waste: but is brought forth out of the nations, and they shall dwell safely all of them. Thou shalt ascend and come like a storm, thou shalt be like a cloud to cover the land, thou and all thy bands, and many people with thee. Thus saith the Lord

God; It shall also come to pass, that at the same time shall things come into thy mind, and thou shalt think an evil thought: And thou shalt say, I will go up to the land of unwalled villages; I will go to them that are at rest, that dwell safely, all of them dwelling without walls, and having neither bars nor gates, to take a spoil, and to take a prey; to turn thy hand upon the desolate places that are now inhabited, and upon the people that are gathered out of the nations, which have gotten cattle and goods, that dwell in the midst of the land. Sheba, and Dedan, and the merchants of Tarshish, with all the young lions thereof, shall say unto thee, Art thou come to take a spoil? Hast thou gathered thy company to take a prey?

To carry away silver and gold, to take away cattle and goods, to take a great spoil? Therefore, son of man, prophesy and say unto Gog, Thus saith the Lord God; In that day when My people of Israel dwelleth safely, shalt thou not know it? And thou shalt come from thy place out of the north parts, thou, and many people with thee, all of them riding upon horses, a great company, and a mighty army: And thou shalt come up against My people of Israel, as a cloud to cover the land; it shall be in the latter days, and I will bring thee against My land, that the heathen may know Me, when I shall be sanctified in thee, O Gog, before their eyes. Thus saith the Lord God; Art thou he of whom I have spoken in old time by My servants the prophets of Israel,

which prophesied in those days many years

that I would bring thee against them? And it

shall come to pass at the same time when

Gog shall come against the land of Israel,

saith the Lord God, that My fury shall

come up in My face. For in My jealousy

and in the fire of My wrath have I spoken.

Surely in that day there shall be a great

shaking in the land of Israel; So that the

fishes of the sea, and the fowls of the

heaven, and the beasts of the field, and all

creeping things that creep upon the earth,

and all the men that are upon the face of the

earth, shall shake at My presence, and the

mountains shall be thrown down, and the

steep places shall fall, and every wall shall

fall to the ground. And I will call for a sword against him throughout all My mountains, saith the Lord God: every man's sword shall be against his brother. And I will plead against him with pestilence and with blood; and I will rain upon him, and upon his bands, and upon the many people that are with him, an overflowing rain, and great hailstones, fire and brimstone. Thus will I magnify Myself, and sanctify Myself; and I will be known in the eyes of many nations, and they shall know that I am the Lord." (Ezekiel 38:1-23 587 B.C. KJV)

"And it shall come to pass in that day, that I will give unto Gog a place there of graves in Israel, the valley of the passengers on the east of the sea: and it shall stop

the noses of the passengers: and there shall
they bury Gog and all his multitude: and
they shall call it the valley of Hamon-gog.
And seven months shall the house of Israel
be burying of them, that they may cleanse
the land. Yea, all the people of the land
shall bury them; and it shall be to them a
renown the day that I shall be glorified,
saith the Lord God. And they shall sever
out men of continual employment, passing
through the land to bury with the passengers those that remain upon the face of the
earth, to cleanse it: after the end of seven
months shall they search. And the passengers that pass through the land, when any
seeth a man's bone, then shall he set up a
sign by it, till the buriers have buried it in

the valley of Hamon-gog." (Ezekiel 39:11-15 587 B.C. KJV)

"Behold, I will make Jerusalem a cup of trembling unto all people round about, when they shall be in the siege both against Judah and against Jerusalem. And in that day will I make Jerusalem a burdensome stone for all people: all that burden themselves with it shall be cut in pieces, though all the people of the earth be gathered together against it." (Zechariah 12:2-3 487 B.C. KJV)

"And this shall be the plague wherewith the Lord will smite all the people that have fought against Jerusalem; Their flesh shall consume away while they stand upon their feet, and their eyes shall consume away in their holes, and their tongue shall consume away in their mouth." (Zechariah 14:12 487 B.C. KJV)

The battle described in the foregoing Scriptures is *not* the Battle of Armageddon which will take place at the end of "the great tribulation period" **just prior** to Christ's return at which time **He destroys all the armies gathered at Armageddon.**

During his *Patmos vision*, the Apostle John sees the "rapture of faithful Christians;" false peace wickedly orchestrated by Satan's Anti-Christ; the rise of the Anti-Christ's false prophet; the nuclear war between the nations; global famine and pestilence; the decimation of the Islamic armies seeking to overthrow

Satan's Anti-Christ; the attack upon Israel
by a federation of nations led by Russia;
the ministry of the "two witnesses" who
defy Anti-Christ and call down from God
numerous plagues upon earth; the awesome
sequential outpouring of God's wrath upon
the kingdom of Anti-Christ; the ministry of
the 144,000 Jewish evangelists during "the
great tribulation period;" the Battle of
Armageddon; the return of Jesus Christ to
Planet Earth; the extermination of the
armies gathered at Armageddon; the
consignment of the Anti-Christ and his
false prophet to the lake of fire; the
imprisonment of Satan for 1,000 years; the
millennium reign of Christ on Earth; the
loosing of Satan after 1,000 years; the final

battle between Jesus Christ and Satan; the consignment of Satan to the lake of fire; the judgment of those who died in their sins after rejecting Christ; the renovation of the heavens and Earth; the descending of the "New Jerusalem" from heaven; and a preview of the eternal Kingdom of God.

John's Patmos vision is triggered by personal communication with Jesus Christ in His glorified form; and with Jesus dictating to John seven letters directed to seven churches located in Asia. Thereafter, John sees Jesus in the form of a slain sacrificial lamb opening a scroll with writing within and without sealed with seven seals. The scene is *before the throne of Almighty God* where a great host of angels are in attendance along with four living creatures (*Seraphims, angels of praise*) and twenty-four Elders (*twelve tribes and twelve apostles*):

"And I saw when the Lamb opened one of the seals, and I heard, as it were the noise of thunder, one of the living creatures saying, Come and see. And I saw, and behold, a white horse: and he that sat on him had a bow; and a crown was given unto him: and he went forth conquering, and to conquer *(the Anti-Christ rising to power through a false peace).* And when He had opened the second seal, I heard the second living creature say, Come and see. And there went out another horse that was red: and power was given to him that sat thereon to take peace from the earth, and that they should kill one another; and there was given unto him a great sword *(the war between the nations and Arab attack upon Israel)*. And when He had opened the third seal, I heard the third living creature say, Come and see. And I beheld, and lo a black horse; and he that sat on him had a pair of balances in his hand. And I heard a voice in the midst of the four living creatures say, A measure of wheat for a penny, and three measures of barley for a penny; and see thou hurt not the oil and the wine *(the unprecedented pollution of earth's soils brings about mass starvation and pestilence while the Anti-Christ and his security forces live like royalty)*. And when

He had opened the fourth seal, I heard the voice of the fourth living creature say, Come and see. And I looked, and behold a pale horse: and his name that sat on him was Death, and Hell followed with him. And power was given unto them over the fourth part of the earth, to kill with sword, and with hunger, and with death, and with the beasts of the earth (*Anti-Christ directs his supernatural powers to destroying all humanity; especially the nation of Israel*). And when He had opened the fifth seal, I saw under the altar the souls of them that were slain for the Word of God, and for the testimony which they held: And they cried with a loud voice, saying, How long, O Lord, holy and true, doest Thou not judge and avenge our blood on them that dwell on the earth? And white robes were given unto every one of them; and it was told unto them, that they should rest yet for a little season, until their fellow servants and their brethren, that should be killed as they were should be fulfilled *(the hunting, torture and murder of "the divine vomit" spoken of by Jesus in His letter to the church members at Laodicea).* And I beheld when He had opened the sixth seal, and, lo, there was a great earthquake: and the sun became black as sackcloth of hair, and the moon became as blood; and the stars of heaven fell unto the earth, even as a fig tree casteth her untimely figs, when she is shaken of a mighty wind. And the heaven departed as a scroll when it is rolled together; and every mountain and island were moved out of their places. And the kings of the earth, and the great men, and the rich men, and the chief captains, and the mighty men, and every bondman, and every free man, hid themselves in the dens and in the rocks of the mountains; And said to the mountains and rocks, Fall on us, and hide us from the face of Him that sitteth on the throne, and from the wrath of the Lamb" For the great day of His wrath is come; and who shall be able to stand? *(those remaining on Earth are given a glimpse into heaven and the outpouring of God's wrath that has begun).* And after these things I saw four angels standing on the four corners of the earth, holding the four winds of the earth, that the wind should not blow on the earth, nor on the sea, nor on any tree. And I saw another angel ascending from the east, having the seal of the Living God: and he cried with a loud voice to the four angels, to whom it was given to hurt the earth and the sea, saying, Hurt not the earth, neither the sea nor the trees, till we have sealed the servants of our God in their foreheads. And I heard the number of them which were sealed: and there were sealed an hundred and forty and four thousand of all the tribes of the children of Israel (*twelve thousand from each tribe to preach the final word of God to humanity).* [Revelation 6:1 to 7:4 96 A.D. KJV)

After seeing the sealing of the 144,000 Jewish evangelists; and the great multitude of **martyred** lukewarm Christians who were **left behind** at the rapture of faithful Christians but knowing enough to refuse to worship the Anti-Christ; John sees the Lamb open the seventh seal:

"And when He had opened the seventh seal, there was silence in heaven about the space of half an hour. And I saw seven angels which stood before God; and to them were given seven trumpets. And another angel came and stood at the altar, having a golden censer; and there was given unto him much incense, that he should offer it with the prayers of all saints upon the golden altar which was before the throne. And the smoke of the incense, which came with the prayers of the saints, ascended up before God out of the angel's hand. And the angel took the censer, and filled it with fire of the altar, and cast it unto the earth: and there were voices, and thundering, and lightening, and an earth-quake; and the seven angels which had the seven trumpets prepared themselves to sound." (Revelation 8:1-6 96 A.D. KJV)

As the seven angels blow their trumpets in sequence, John sees more of God's wrath descend upon Anti-Christ and his earthly kingdom. Trumpets one through five introduce hail and fire mingled with blood; one third of trees burning and all green grass burning; toxic masses from interstellar space destroying ships; poisoning the oceans and killing marine life; a decrease of one third in sunlight, moonlight and starlight; and hellish locusts released from the abyss whose sting torments for five months without relief through death.

When the sixth trumpet sounds, the Islamic hordes set out to exterminate "all the infidels" and two and a third billion people on earth die including one and three quarter billion of the Islamic attackers during the jihad: when the Islamic murderers face over two hundred million orientals armed with weapons of mass destruction.

Thereafter, John sees a mighty angel standing with one foot upon the earth and one foot upon the sea while lifting up his hand to heaven and swearing by Him who lives forever that there will be time no longer but that the mystery of God, as He has declared to His prophets, is finished.

Then, the "two witnesses" call down from heaven more plagues upon the kingdom of Anti-Christ before being killed and then resurrected before the eyes of all the world during a devastating earthquake *(instantaneous world news coverage made possible by satellite TV).*

The Anti-Christ **(***Satan in the flesh)* along with his companion false prophet (*the world leader of apostate Christianity)* enforce "the mark of the beast" without which no one can buy or sell. The mark is designed to identify all who refuse to worship the Anti-Christ and all who refuse are tortured and beheaded.

The Anti-Christ declared himself to be God Almighty, casts away his false prophet, destroys the capital city of apostate Christianity and sets up his earthly throne inside the Jewish Temple in Jerusalem while at the same time setting out to exterminate all Jews.

All of the apocalyptic events John sees in his Patmos vision occur within a seven year period beginning with the signing of the **"end time" peace treaty between the Anti-Christ and the nation**

of Israel. The **"end times prophetic time window"** opened in 1948 A.D. when Israel was reborn as an independent sovereign nation. In order to follow the **timing of the Patmos vision**, it is essential to distinguish between the end times prophetic time window and the seven year reign of the Anti-Christ.

The sounding of the seventh trumpet signals the total destruction of the kingdom of the Anti-Christ culminating in the Battle of Armageddon and the return of Jesus Christ to Planet Earth. The nations still remaining upon earth during the reign of the Anti-Christ, as the seven year period comes to a close, rebel against the lies and treachery perpetrated upon them by Satan in the flesh. The forces defending the Anti-Christ establish a defensive perimeter in the valley of Megiddo in the plain of Jezreel when the attacking armies cross the Euphrates thereby causing Israel to be encircled by hostile forces numbering roughly half a billion.

As the stage is being set for the Battle of Armageddon, seven additional vials of the wrath of God are poured out upon Earth: a noisome and grievous sore falls upon all who worshiped Anti-Christ; the seas turns to blood killing everything living in the earth's oceans; all the fresh waters on Earth also turn to blood; Earth orbits closer to the sun scorching the planet with great heat; the solar heat dissipates and thick darkness covers Earth causing people to gnaw their tongue pursuant to frigid temperatures; the Euphrates dries up providing a quick route for the Oriental armies attacking the Anti-Christ; and a mighty earthquake is followed by hail-stones weighing eighty to one hundred pounds each.

John sees the Battle of Armageddon interrupted by the appearance of Jesus Christ in preparation for His millennium reign on Earth:

"And the beast (*Anti-Christ*) was taken, and with him the false prophet that wrought miracles before him, with which he had deceived them that had received the mark of the beast, and them that worshiped his image. These both were cast alive into a lake of fire burning with brimstone. And the remnant were slain with the sword (*the spoken word*) of Him that sat upon the horse, which sword proceeded out of His mouth: and all the fowls were filled with their flesh." (Revelation 19:20-21 96 A.D. KJV)

Satan, himself, is bound and imprisoned within the "bottomless pit" during the millennium reign of Jesus Christ on Earth from His throne in Jerusalem. The remnants of the nations and the Jews who survived Satan's murderous onslaught will enter the millennium reign in their natural bodies and reproduce during the 1,000 years. Death will remain an enemy but one who dies a hundred years old will be considered a child:

"But be ye glad and rejoice forever in that which I create: for behold, I create in Jerusalem a rejoicing, and her people a joy. And I will rejoice in Jerusalem, and joy in My people: and the voice of weeping shall be no more heard in her, nor the voice of crying. There shall be no more thence an infant of days, nor an old man that hath not fulfilled his days: for the child shall die a hundred years old; but the sinner being a hundred years old shall be accursed." (Isaiah 65:18-20 698 B.C. KJV)

"And it shall come to pass, that every one that is left of **all the nations** which came against Jerusalem shall even go up from year to year to worship the King, the Lord of Hosts, and to keep the feast of tabernacles. And it shall be, that whoso will not come up of all the families of the earth unto Jerusalem to worship the King, the Lord of Hosts, even upon them shall be no rain." (Zechariah 14:16-17 487 B.C. KJV)

"And when the thousand years are expired, Satan shall be loosed out of his prison, and shall go out to deceive the nations which are in the four quarters of the earth, Gog and Magog, to gather them together to battle: the number of whom is as the sand of the sea. And they went up on the breadth of the earth, and compassed the camp of the saints about, and the beloved city: and fire came down from God out of

heaven, and devoured them. And the devil
that deceived them was cast into the lake of
fire and brimstone, where the beast and
false prophet are, and shall be tormented
day and night for ever and ever. And I saw
a great white throne, and Him that sat on it,
from whose face the earth and the heaven
fled away; and there was found no place for
them. And I saw the dead, small and great,
stand before God; and the books were
opened: and another book was opened,
which is the book of life: and the dead were
judged out of those things which were
written in the books, according to their
works. And the sea gave up the dead which
were in it; and death and hell delivered up
the dead which were in them: and they

were judged every man according to their works. And death and hell were cast into the lake of fire. This is the second death. And whosoever was not found written in the book of life was cast into the lake of fire." (Revelation 20:7-15 96 A.D. KJV)

"And I saw a new heaven and a new earth: for the first heaven and the first earth were passed away; and there was no more sea. And I John saw the holy city, New Jerusalem, coming down from God out of heaven, prepared as a bride adorned for her husband. And I heard a great voice out of heaven saying, Behold, the tabernacle of God is with men, and He shall dwell with them, and they shall be His people, and God Himself shall be with them and be their God. And God shall wipe away all tears from their eyes; and there shall be no more death, neither sorrow, nor crying, neither shall there be any more pain: for the former things are passed away. And He that sat upon the throne said, Behold, I make all things new. And He said unto me, Write: for these words are true and faithful." (Revelation 21:1-5 96 A.D. KJV)

"But as it is written, Eye hath not seen, nor ear heard, neither has it entered into the heart of man, the things God has prepared for them that love Him." (I Corinthians 2:9 59 A.D. KJV)

The cost of the cross was much too heavy to be calculated by human standards. It is more or less comparable to trying to calculate ten to the billionth power. Yet, that cost **was balanced** to the sin debt of humanity. We owed a debt we could not pay; Christ Jesus paid a debt He did not owe; **because we needed someone to wash our sins away.** Those who wind up in the lake of fire will be there only because they insisted upon their right to spend eternity with their spiritual father.

Epilogue

He had been without frontal vision for years but had learned to cope with limited peripheral sight. During the last three years, his quality of life had diminished to commuting slowly from his bed to the kitchen table. The cartilage in both knees had thinned allowing bone to grind against bone. He gasped for breath when hobbling to the bathroom and his bowels became more irregular. He suffered cramps in his hands, legs and feet and his skin had turned a bluish black. He depended on nine prescription drugs and

three inhalers to maintain his fragile
pulmonary and cardiac functions. His
hearing aids and dentures became more and
more useless.

Early one morning, he fell in his kitchen
and could not get up again. His caretaker
could not lift him. An ambulance came and
took him to his death-bed. He was a mere
child in years having seen only ninety-five
winters. He had been born with an immortal soul but with a mortal body. His death
had been appointed for sixty centuries and
now the sentence was being carried out.
William V. Alexander was returning to the
dust from which his ancestors had been
created approximately forty centuries

before the birth of Jesus Christ. The
hospital room was stark, quiet and smelled
of death. William's breathing was barely
perceptible as his lungs struggled against
pneumatic fluid. His kid-neys failed and his
eyes turned upward showing only slits of
white. He came into the world with nothing
and he was leaving with nothing. He
struggled to hold on to his flesh with every
last beat of his heart; and then he was gone.
His body would soon be laid to rest with
the powerful of the earth, the wise, the
good, aborted fetuses, mutilated children,
drunks, harlots, whore mongers, murderers,
homosexuals, atheists, fools, self-appointed
wise men, those who believed in Jesus
Christ and those who did not. His life was

like a vapor that appeared for a little while
and then vanished. He had a body, a spirit
and a soul. His body was like every other
body in that it required daily nurturing and
began to die from birth. His spirit consisted
of the breath of God which He breathed
into Adam's nostrils causing Adam to
become a living soul. His spirit maintained
God consciousness thereby distinguishing
between good and evil. His soul had
developed into his innermost seat of
emotions and desires, having been shaped
by personal decisions which now followed
him into eternity. William V. Alexander
believed in Jesus Christ and was heading
into everlasting life where mortal regains
immortality -- that which was lost in the

Garden of Eden …..the redemption for
which God sacrificed even Himself

Bibliography

Authorized King James Version, Holy
Bible, 1611 A.D.; public domain with
minor paraphrasing by author.
Carbon-14 Dating, Radiometric Dating and
Tree Ring DatingPlastino, W.; Kaih^ola,
L.; Bartolomei, P.; Bella, F. (2001)."Cosmic
Background Reduction In The Radiocarbon
Measurement By ScintillationSpectrometry
At The Underground Laboratory Of Gran
Sasso". *Radiocarbon* **43** (2A): 157–
161. https://digitalcommons.library.arizona.
edu/objectviewer?o=http%3A%2F

%2Fradiocarbon.library.arizona.edu %2Fvolume43%2Fnumber2A %2Fazu_radiocarbon_v43_n2a_157_161_v .pdf. ^ Arnold, J. R.; Libby, W. F. (1949). "Age Determinations by Radiocarbon Content: Checks with Samples of Known Age". *Science* **110** (2869): 678–680. doi:10.1126/science.110.2869.678. PMID 15407879.http://hbar.phys.msu.ru/gorm/fo menko/libby.htm. ^Willard Frank Libby

1. ^ *a* *b* *c* Münnich KO, Östlund HG, de Vries H (1958). "Carbon-14 Activity during the past 5,000 Years". *Nature* **182** (4647): 1432–3. doi:10.1038/1821432a0.

^ *a* *b* Ramsey, C. Bronk (2008). "Radiocarbon dating: revolutions in understanding".*Archaeometry* **50** (2):24d

∧ Scott, EM (2003). "The Fourth International Radiocarbon Intercomparison (FIRI).". *Radiocarbon* **45**: 135–285.

∧ *a b* "NOSAMS Radiocarbon Data and Calculations". Woods Hole Oceanographic Institution. http://www.nosams.whoi.edu/clients/data.html. ∧ Taylor RE, Southon J (2007). "Use of natural diamonds to monitor ^{14}C AMS instrument backgrounds". *Nuclear Instruments and Methods in Physics Research B* **259**: 282–28. doi:10.1016/j.nimb.2007.01.239.

∧ Stuiver M, Reimer PJ, Braziunas TF (1998). "High-precision radiocarbon age calibration for terrestrial and marine

samples". *Radiocarbon* **40**: 1127–
51. http://depts.washington.edu/qil/datasets/
uwten98_14c.txt.

^ "Atmospheric $\delta^{14}C$ record
from Wellington". Carbon Dioxide
Information Analysis
Center. http://cdiac.esd.ornl.gov/trends/co2/
welling.html. Retrieved 1 May

2008. ^ "$\delta^{14}CO_2$ record from
Vermunt". *Carbon Dioxide Information
Analysis Center.*
http://cdiac.esd.ornl.gov/trends/co2/cent-
verm.html. Retrieved 1 May
2008. ^ "Radiocarbon dating". Utrecht
University. http://www1.phys.uu.nl/ams/Ra
diocarbon.htm. Retrieved 1 May

2008. ^ Kudela K. and Bobik P. (2004). "Long-Term Variations of Geomagnetic Rigidity Cutoffs". Solar Physics **224**: 423–431. doi:10.1007/s11207-005-6498-9.

^ Reimer, Paula J.; Brown, Thomas A.; Reimer, Ron W. (2004). "Discussion: Reporting and Calibration of Post-Bomb ^{14}C Data". *Radiocarbon* **46** (3): 1299–1304

^ These results were obtained from a Monte Carlo analysis calibrating simulated measurements of varying precision using the 1993 version of the calibration curve. The width of the uncertainty represents a 2σ uncertainty (that is, a likelihood of 95% that the date appears between these limits). Niklaus TR,

Bonani G, Suter M, Wölfli W (1994).
"Systematic investigation of uncertainties
in radiocarbon dating due to fluctuations in
the calibration curve". *Nuclear Instruments
and Methods in Physics Research* **92**: 194–
200. doi:10.1016/0168-583X(94)96004-6.
^ Reimer Paula J *et al.* (2004).
"INTCAL04 Terrestrial Radiocarbon Age
Calibration, 0–26 Cal Kyr
BP". *Radiocarbon* **46** (3): 1029–
1058. http://digitalcommons.library.arizona
.edu/objectviewer?
o=http://radiocarbon.library.arizona.edu/Vo
lume46/Number3/azu_radiocarbon_v46_n3
_1029_1058_v.pdf. A web interface
is here. ^ Reimer, P.J.; et. al.
(2009). "IntCal09 and Marine09

Radiocarbon Age Calibration Curves, 0–50,000 Years cal BP". *Radiocarbon* **51** (4): 1111–1150. http://researchcommons.waikato.ac.nz/bitstream/10289/3622/1/Hogg%20Intcal09%20and%20Marine09.pdf. ^ Balter, Michael (15 Jan 2010). "Radiocarbon Daters Tune Up Their Time Machine". *ScienceNOW Daily News*. http://sciencenow.sciencemag.org/cgi/content/full/2010/115/3.
^ Godwin, H. (1962). "Half-life of Radiocarbon". *Nature* **195** (4845): 984. doi:10.1038/195984a0. ^ Libby WF (1955). *Radiocarbon dating* (2nd ed.). Chicago: University of Chicago Press. ^ Lerman, J. C.; Mook, W. G.; Vogel, J. C.; de Waard, H. (1969). "Carbon-

14 in Patagonian Tree
Rings". *Science* **165** (3898): 1123–
1125. doi:10.1126/science.165.3898.1123.
PMID 17779805.

^ McNichol AP, Schneider RJ, von Reden
KF, Gagnon AR, Elder KL, NOSAMS, Key
RM, Quay PD (October 2000). "Ten years
after - The WOCE AMS radiocarbon
program". *Nuclear Instruments and
Methods in Physics Research, Section B:
Beam Interactions with Materials and
Atoms* **172** (1-4): 479–
84. doi:10.1016/S0168-583X(00)00093-8.

^ Stuiver M, Braziunas TF (1993).
"Modelling atmospheric ^{14}C influences
and ^{14}C ages of marine samples to 10,000

BC". *Radiocarbon* **35** (1): 137.

^ *a b* Kolchin BA, Shez YA
(1972). *Absolute archaeological datings
and their problems*. Moscow:
Nauka. ^ Crowe C (1958). "Carbon-14
activity during the past 5000
years". *Nature* **182** (4633): 470–
1. doi:10.1038/182470a0. ^ Barker H
(1958). "Carbon-14 Activity during the past
5,000 Years". *Nature* **182** (4647):
1433. doi:10.1038/1821433a0. ^ Libby
WF (1962). "Radiocarbon; an atomic
clock". *Annual Science and Humanity
Journal*. ^ Wang YJ; Cheng, H; Edwards,
RL; An, ZS; Wu, JY; Shen, CC; Dorale, JA
(2001). "A High-Resolution Absolute-
Dated Late Pleistocene Monsoon Record

from Hulu Cave, China.". *Science* **294** (5550): 2345–2348.doi:10.1126/science.1064618. PMID 11743199. ^ Beck JW; Richards, DA; Edwards, RL; Silverman, BW; Smart, PL; Donahue, DJ; Hererra-Osterheld, S; Burr, GS et al. (2001). "Extremely large variations of atmospheric C-14 concentration during the last glacial period.". *Science* **292** (5526): 2453–2458. doi:10.1126/science.1056649. PMID 11349137. ^ *a* *b* Hoffmann DL; Beck, J. Warren; Richards, David A.; Smart, Peter L.; Singarayer, Joy S.; Ketchmark, Tricia; Hawkesworth, Chris J. (2010). "Towards radiocarbon calibration beyond 28 ka using speleothems from the Bahamas".*Earth and*

Planetary Science Letters **289**: 1–10. Bibcode 2010E&PSL.289....1H. doi:10.1016/j.epsl.2009.10.004.

 ^ Jensen MN (2001). "Peering deep into the past". University of Arizona, Department of Physics. http://www.physics.arizona.edu/physics/public/beck-citizen.html. ^ Pennicott K (10 May 2001). "Carbon clock could show the wrong time" *PhysicsWeb*. http://physicsworld.com/cws/article/news/2676. Big Bang Theory ^ D. N. Spergel et al. (2007). "Three-Year Wilkinson Microwave Anisotropy Probe (WMAP) Observations: Implications for Cosmology".*Astrophysical Journal Supplement Series* **170** (2): 377–

408. arXiv:astro-ph/0603449. Bibcod
e2007ApJS..170..377S. doi:10.1086/51370
0. ^ *a b* Dodelson, Scott (2003). *Modern
Cosmology*. Academic Press. ISBN 0-12-
219141-2. ^ *a b* Liddle, Andrew; David
Lyth (2000). *Cosmological Inflation and
Large-Scale Structure*. Cambridge. ISBN 0-
521-57598-2. ^ *a b* Padmanabhan, T.
(1993). *Structure formation in the universe*.
Cambridge University Press. ISBN 0-521-
42486-0. ^ Peebles, P. J. E. (1980). *The
Large-Scale Structure of the Universe*.
Princeton University Press. ISBN 0-691-
08240-5. ^ Kolb, Edward; Michael Turner
(1988). *The Early Universe*. Addison-
Wesley. ISBN 0-201-11604-9. ^ Wayne

Hu and Scott Dodelson (2002). "Cosmic microwave background anisotropies". *Ann. Rev. Astron. Astrophys.* **40** (1): 171–216. arXiv:astro-ph/0110414. Bibcode 2002ARA&A..40..171H. doi:10.1146/annurev.astro.40.060401.093926.

^ *a* *b* Edmund Bertschinger (1998). "Simulations of structure formation in the universe". *Annual Review of Astronomy and Astrophysics* **36** (1): 599–654. Bibcode 1998ARA&A..36..599B. doi:10.1146/annurev.astro.36.1.599. ^ Harrison, E. R. (1970). "Fluctuations at the threshold of classical cosmology". *Phys. Rev.* **D1**: 2726. Bibcode 1970PhRvD...1.2726H. doi:10.1103/PhysRevD.1.2726. ^ Peebles, P. J. E.; Yu, J. T.

(1970). "Primeval adiabatic perturbation in an expanding universe". *Astrophysical Journal* **162**:

815. Bibcode 1970ApJ...162..815P. doi:10.1086/150713. ^ Ya; Zel'dovich, B. (1972). "A hypothesis, unifying the structure and entropy of the universe". *Monthly Notices of the Royal Astronomical Society* **160**. Bibcode 1972MNRAS.160P...1Z. ^ R. A. Sunyaev, "Fluctuations of the microwave background radiation," in *Large Scale Structure of the Universe* ed. M. S. Longair and J. Einasto, 393. Dordrecht: Reidel 1978. ^ U. Seljak and M. Zaldarriaga (1996). "A line-of-sight integration approach to cosmic microwave background anisotropies". *Astrophysics*

J. **469**: 437–444. arXiv:astro-ph/9603033.
Bibcode 1996ApJ...469..437S. doi:10.1086/177793. ^ Springel, V. *et al* (2005).
"Simulations of the formation, evolution and clustering of galaxies and quasars". *Nature* **435** (7042): 629–636. arXiv:astro-ph/0504097. Bibcode 2005Natur.435..629S. doi:10.1038/nature03597.PMID 15931216.
Quantum Mechanics ^ Richard P. Feynman, *QED*, p. 10 ^ Landau, L. D.; E. M. Lifshitz (1996). *Statistical Physics* (3rd Edition Part 1 ed.). Oxford: Butterworth-Heinemann. ISBN 0521653142. ^ This was published (in German) as Planck, Max (1901). "Ueber das Gesetz der Energieverteilung im Normalspectrum".

Ann. Phys. **309** (3): 553–63. Bibcode 1901AnP...309..553P. doi:10.1002/andp.19013090310. http://www.physik.uniaugsburg.de/annalen/history/historic-papers/1901_309_553-563.pdf . English translation: "On the Law of Distribution of Energy in the Normal Spectrum". ^ Francis Weston Sears (1958). *Mechanics, Wave Motion, and Heat*. Addison-Wesley. p. 537. http://books.google.com/books?hl=en&q=%22Mechanics%2C+Wave+Motion%2C+and+Heat%22+%22where+n+%3D+1%2C%22&btnG=Search+Books. ^ "The Nobel Prize in Physics 1918". The Nobel Foundation. http://nobelprize.org/nobel_pri

zes/physics/laureates/1918/. Retrieved 2009-08-01. ^ Kragh, Helge (1 December 2000). "Max Planck: the reluctant revolutionary".

PhysicsWorld.com. http://physicsworld.co m/cws/article/print/373 ^ Einstein, Albert (1905). "Über einen die Erzeugung und Verwandlung des Lichtes betreffenden heuristischen Gesichtspunkt". *Annalen der Physik* **17**: 132–148. Bibcode 1905AnP...322..132E.oi:10.1002/andp.1905 3220607.http://www.zbp.univie.ac.at/doku mente/einstein1.pdf. , translated into English as On a Heuristic Viewpoint Concerning the Production and Transformation of Light. The term "photon" was introduced in 1926.^ Taylor,

J. R.; Zafiratos, C. D.; Dubson, M. A.
(2004). *Modern Physics for Scientists and
Engineers*. Prentice Hall. pp. 127–
9. ISBN 0135897890. ^ Dicke and
Wittke, *Introduction to Quantum
Mechanics*, p. 12 ^
http://ntrs.nasa.gov/archive/nasa/casi.ntrs.n
asa.gov/19680009569_1968009569.pdf*ab*T
aylor, J. R.; Zafiratos, C. D.; Dubson, M. A.
(2004). *Modern Physics for Scientists and
Engineers*. Prentice Hall. pp. 147–
8. ISBN 0135897890. ^ McEvoy, J. P.;
Zarate, O. (2004). *Introducing Quantum
Theory*. Totem Books. pp. 70–89, especially
p. 89. ISBN 1840465778. ^ *World Book
Encyclopedia*, page 6, 2007. ^ Dicke and
Wittke, *Introduction to Quantum*

Mechanics, p. 10f. ^ J. P. McEvoy and Oscar Zarate (2004). *Introducing Quantum Theory*. Totem Books. p. 110f. ISBN 1-84046-577-8. ^ Aezel, Amir D., *Entanglrment*, p. 51f. (Penguin, 2003) ISBN 0-452-28457 ^ J. P. McEvoy and Oscar Zarate (2004). *Introducing Quantum Theory*. Totem Books. p. 114. ISBN 1-84046-577-8. ^ Heisenberg's Nobel Prize citation ^ W. Moore, *Schrödinger: Life and Thought*, Cambridge University Press (1989), p. 222. ^ Heisenberg first published his work on the uncertainty principle in the leading German physics journal *Zeitschrift für Physik*: Heisenberg, W. (1927). "Über den anschaulichen Inhalt der quantentheoretischen Kinematik und

Mechanik". *Z. Phys.* **43** (3–4): 172–198. Bibcode 1927ZPhy...43..172H. doi:10.1007/BF01397280. ^ Nobel Prize in Physics presentation speech, 1932 ^ *a b* Linus Pauling, **The Nature of the Chemical Bond**, p. 47 ^ E. Schrödinger, *Proceedings of the Cambridge Philosophical Society*, 31 (1935), p. 555says: "When two systems, of which we know the states by their respective representation, enter into a temporary physical interaction due to known forces between them and when after a time of mutual influence the systems separate again, then they can no longer be described as before, viz., by endowing each of them with a representative of its own. I

would not call that *one* but rather *the*
characteristic trait of quantum
mechanics."^ "Quantum Nonlocality and
the Possibility of Superluminal Effects",
John G. Cramer,
http://www.npl.washington.edu/npl/int_rep/
qm_nl.htmlTheory of Relativity ^ Einstein
A. (1916 (translation 1920)), *Relativity:*
The Special and General Theory, New
York: H. Holt and Company ^ Planck,
Max (1906), "The Measurements of
Kaufmann on the Deflectability of β-Rays
in their Importance for the Dynamics of the
Electrons", *Physikalische Zeitschrift* **7**:
753–761 ^ Miller, Arthur I. (1981), *Albert*
Einstein's special theory of relativity.
Emergence (1905) and early interpretation

(1905–1911), Reading: Addison–Wesley, ISBN 0-201-04679-2 ^ *a b c d e f g* Will, Clifford M (August 1, 2010). "Relativity". *Grolier Multimedia Encyclopedia*. http://gme.grolier.com/articl e?assetid=0244990-0. Retrieved 2010-08-01. ^ *a b c* Will, Clifford M (August 1, 2010). "Space-Time Continuum". *Grolier MultimediaEncyclopedia*. http://gme.grolier .com/article?assetid=0272730-0. Retrieved 2010-08-01. ^ *a b c* Will, Clifford M (August 1, 2010). "Fitzgerald-Lorentz contraction". *Grolier Multimedia Encyclopedia*. http://gme.grolier.com/articl e?assetid=0107090-0. Retrieved 2010-08-01. ^ *a b c d* Einstein's letter to the London

Times in 1919.Einstein Albert (Nov. 28, 1919). "What is the theory of relativity?"". *The London Times*: pp. 4.

1."Age of the Earth". U.S. Geological Survey. 1997. 2006-01-10.

1.**Jump up^** Manhesa, Gérard; Allègre, Claude J.; Dupréa, Bernard & Hamelin, Bruno (1980). "Lead isotope study of basic-ultrabasic layered complexes: Speculations about the age of the earth and primitive mantle characteristics". *Earth and Planetary Science Letters***47** (3): 370–382. Bibcode:1980E&PSL..47..370M. doi:10.10 16/0012-821X(80)90024-2.

2.^Jump up to:*a b c* Boltwood, B. B. (1907). "On the ultimate disintegration products of the radio-active elements. Part II. The disintegration products of uranium". *American Journal of Science* **23** (134):77-88. doi:10.2475/ajs.s4-23.134.78. For the abstract, see: Chemical Abstracts Service, American Chemical Society (1907). *Chemical Abstracts*. New York, London: American Chemical Society. p. 817. Retrieved2008-12-19.

3.**Jump up**^ Wilde, S. A.; Valley, J. W.; Peck, W. H.; Graham C. M. (2001-01-11). "Evidence from *detrital zircons* for the existence of *continental crust* and oceans on the Earth 4.4 Gyr ago". *Nature* **409** (6817): 175–178. doi:10.1038/35051550. *PMID 11196637*.

4.**Jump up**^ Valley, John W.; Peck,

William H.; Kin, Elizabeth M.
(1999). "Zircons Are Forever"(PDF). *The Outcrop, Geology Alumni Newsletter.* University of Wisconsin-Madison. pp. 34–35. Retrieved 2008-12-22.

5.**Jump up**^ Wyche, S.; Nelson, D. R.; Riganti, A. (2004). "4350–3130 Ma detrital zircons in the Southern Cross Granite–Greenstone Terrane, Western Australia: implications for the early evolution of the Yilgarn Craton". *Australian Journal of Earth Sciences* **51** (1): 31–45. doi:10.1046/j.1400-0952.2003.01042.x.

6.**Jump up**^ Amelin, Y; Krot, An; Hutcheon, Id; Ulyanov, Aa (Sept. 2002). . *Science* **297** (5587): 1678–83.Bibcode:2002Sci...297.1678A. doi:10.1

126/science.1073950. ISSN 0036-8075.PMID 12215641.

7.**Jump up**^ Baker, J.; Bizzarro, M.; Wittig, N.; Connelly, J.; et al. (2005-08-25). "Early planetesimal melting from an age of 4.5662 Gyr for differentiated meteorites". *Nature***436** (7054): 1127–1131. Bibcode:2005Natur.436.1127B. doi:10.1038/nature03882.PMID 16121173.

8.**Jump up**^ *Lyell, Charles, Sir* (1866). *Elements of Geology; or, The Ancient Changes of the Earth and its Inhabitants as Illustrated by Geological Monuments* (Sixth ed.). New York: D. Appleton and company. Retrieved 2008-12-19.

9.^ Jump up to:*a b* Stiebing, William H. (1994). *Uncovering the Past*. Oxford

University Press US.ISBN 0-19-508921-9.

10.^ Jump up to:*a b* Brookfield, Michael E. (2004). *Principles of Stratigraphy.* Blackwell Publishing. p. 116. ISBN 1-4051-1164-X.

11.**Jump up**^ Fuller, J. G. C. M. (2007-07-17). "Smith's other debt, John Strachey, William Smith and the strata of England 1719–1801". *Geoscientist.* The Geological Society.Archived from the original on 24 November 2008. Retrieved 2008-12-19.

12.**Jump up**^ Burchfield, Joe D. (1998). "The age of the Earth and the invention of geological time".*Geological Society, London, Special Publications* **143** (1): 137–143.Bibcode:1998GSLSP.143..137B. doi:10.1144/GSL.SP.1998.143.01.12.

13.^ Jump up to:*a b* England, P.; Molnar, P.; Righter, F. (January 2007). "John Perry's neglected critique of Kelvin's age for the Earth: A missed opportunity in geodynamics". *GSA Today* **17** (1): 4–9. doi:10.1130/GSAT01701A.1.

14.Jump up^ Dalrymple (1994) pp. 14–17, 38

15.**Jump up**^ Borenstein, Seth (November 13, 2013). "Oldest fossil found: Meet your microbial mom". *Excite* (Yonkers, NY: Mindspark Interactive Network). Associated Press. Retrieved 2015-03-02.)

16.^ Jump up to:*a b c* Dalrymple (1994) pp. 14–17

17.**Jump up**^ Paul J. Nahin (1985) Oliver Heaviside, Fractional Operators, and the

Age of the Earth, IEEE Transactions on Education E-28(2): 94–104, link from IEEE Explore

18.**Jump up**^ Dalrymple (1994) pp. 14, 43

19.^ Jump up to:*a b c* Nichols, Gary (2009). "21.2 Radiometric Dating". *Sedimentology and Stratigraphy*. John Wiley & Sons. pp. 325–327. ISBN 978-1405193795.^ Jum p up to:*a b* England, Philip C.; Molnar, Peter; Richter, Frank M. (2007). "Kelvin, Perry and the Age of the Earth". *American Scientist* **95** (4): 342–349. doi:10.1511/2007.66.3755."The Dancing Wu Li Masters" by Gary Zukav (Perennial, 1979)"God and the New Physics" by Paul Davies (Penguin, 1983)"A Brief History of Time" by Stephen

Hawking (Bantam, 1988)"Deep Time" by David Darling (Bantam Press, 1989)"The Physics of Star Trek" by Lawrence M. Krauss (Basic Books, 1995)"Just Six Numbers: The Deep Forces That Shape the Universe" by Martin Rees (Basic Books, 2000)"A Briefer History of Time" by Stephen Hawking (Bantam, 2005)"Quantum Theory Cannot Hurt You" by Marcus Chown (Faber and Faber 2007)"The Quirks and Quarks Guide to Space" by Jim Lebans (McClelland and Stewart, 2008)"Death from the Skies! These are the Ways the World Will End..." by Philip Plait (Viking, 2008)"You Are Here - A Portable History of the Universe" by Christopher Potter (Vintage, 2009)"The

Elegant Universe" (Nova, 2003)"The Universe" (History Channel, 2007)"Journey to the Edge of the Universe" (National Geographic Channel, 2008)"Stephen Hawking, Master of the Universe" (Channel 4, 2008)"Into the Universe With Stephen Hawking" (Discovery Channel, 2010) "Through the Wormhole" (Science Channel, 2010)"How the Universe Works" (Discovery Channel, 2010)"The Fabric of the Cosmos" (Nova, 2011)

WEBSITES

- WikiPedia (http://en.wikipedia.org/wiki/)
- NASA: Universe 101 (http://map.gsfc.nasa.gov/universe/)

•BBC: Space(http://www.bbc.co.uk/science/space/)

•How Stuff Works: Astronomy (http://science.howstuffworks.com/astronomy-channel.htm)

•Curious About Astronomy (http://curious.astro.cornell.edu/)

•MacTutor History of Mathematics Archive (http://www-history.mcs.st-andrews.ac.uk/)

•About.com: Physics (http://physics.about.com/)

•Internet Encyclopedia of Science (http://www.daviddarling.info/encyclopedia/ETEmain.html)

•Flash Modern Physics Tutorials

(http://www.gilestv.com/tutorials/tutorials.html)

•Physics 2000
(http://www.colorado.edu/physics/2000/index.pl)

•Physics for Future Presidents
Lectures by Richard A. Muller, U.C.
Berkeley, on YouTube
(http://www.youtube.com/watch?v=6ysbZ_j2xi0)

References[edit]

1. **Jump up**^ "autonomic nervous system" at *Dorland's Medical Dictionary*
2. **Jump up**^ Schmidt, A; Thews, G (1989). "Autonomic Nervous System". In Janig, W. *Human Physiology* (2 ed.).

New York, NY: Springer-Verlag.
pp. 333–370.

3. ^ Jump up to:*a b* Allostatic load
notebook: Parasympathetic Function -
1999, MacArthur research
network, UCSF

4. **Jump up**^ Pocock, Gillian
(2006). *Human Physiology* (3rd ed.).
Oxford University Press. pp. 63–
64. ISBN 978-0-19-856878-0.

5. **Jump up**^ Belvisi, Maria G.; David
Stretton, C.; Yacoub, Magdi; Barnes,
Peter J. (1992). "Nitric oxide is the
endogenous neurotransmitter of
bronchodilator nerves in
humans". *European Journal of
Pharmacology* **210** (2): 221–

2. doi:10.1016/0014-2999(92)90676-U.PMID 1350993.

6. **Jump up**^ Costanzo, Linda S. (2007). *Physiology*. Hagerstwon, MD: Lippincott Williams & Wilkins. p. 37. ISBN 0-7817-7311-3.

7. **Jump up**^ Essential Clinical Anatomy. K.L. Moore & A.M. Agur. Lippincott, 2 ed. 2002. Page 199

8. ^ Jump up to:*a b* Unless specified otherwise in the boxes, the source is: Moore, Keith L.; Agur, A. M. R. (2002). *Essential Clinical Anatomy* (2nd ed.). Lippincott Williams & Wilkins. p. 199.ISBN 978-0-7817-5940-3.

9. **Jump up**^ Neil A. Campbell, Jane B. Reece: Biologie. Spektrum-Verlag

Absence of Chaos   Don Alexander   page 312

Heidelberg-Berlin 2003,ISBN 3-8274-1352-4